Evolution in Isolation
The Search for an Island Syndrome in Plants

Oceanic islands are storehouses for unique creatures. Zoologists have long been fascinated by island animals because they break all the rules. Speedy, nervous, little birds repeatedly evolve to become plump, tame, and flightless on islands. Equally strange and wonderful plants have evolved on islands. However, plants are very poorly understood relative to animals. Do plants repeatedly evolve similar patterns in dispersal ability, size, and defence on islands? This volume answers this question for the first time using a modern quantitative approach. It not only reviews the literature on differences in defence, loss of dispersal, changes in size, alterations to breeding systems, and the loss of fire adaptations, but also brings new data into focus to fill gaps in current understanding. By firmly establishing what is currently known about repeated patterns in the evolution of island plants, this book provides a roadmap for future research.

KEVIN C. BURNS is Professor in Biology at Victoria University of Wellington, New Zealand. He has been a practicing researcher for 15 years and has published over 100 papers in scientific journals, including *Ecology*, *Ecology Letters*, and *Science*. He is fascinated by how organisms evolve on islands and has worked on archipelagos across the globe, including New Caledonia, New Zealand, the Chatham Islands, the California Islands, and Lord Howe Island.

Evolution in Isolation

The Search for an Island
Syndrome in Plants

KEVIN C. BURNS
Victoria University of Wellington

CAMBRIDGE
UNIVERSITY PRESS

CAMBRIDGE
UNIVERSITY PRESS

University Printing House, Cambridge CB2 8BS, United Kingdom

One Liberty Plaza, 20th Floor, New York, NY 10006, USA

477 Williamstown Road, Port Melbourne, VIC 3207, Australia

314-321, 3rd Floor, Plot 3, Splendor Forum, Jasola District Centre, New Delhi - 110025, India

79 Anson Road, #06-04/06, Singapore 079906

Cambridge University Press is part of the University of Cambridge.

It furthers the University's mission by disseminating knowledge in the pursuit of education, learning and research at the highest international levels of excellence.

www.cambridge.org
Information on this title: www.cambridge.org/9781108422017
DOI: 10.1017/9781108379953

© Kevin C. Burns 2019

First published 2019

A catalogue record for this publication is available from the British Library

Library of Congress Cataloging in Publication data
Names: Burns, Kevin C., 1970– author.
Title: Evolution in isolation : the search for an island syndrome in plants /
    Kevin C. Burns, Victoria University of Wellington.
Description: Cambridge, United Kingdom ; New York, NY : Cambridge University
    Press, 2019. | Includes bibliographical references and index.
Identifiers: LCCN 2018046543 | ISBN 9781108422017 (hardback)
Subjects: LCSH: Island plants–Evolution.
Classification: LCC QK938.I84 B87 2019 | DDC 578.75/2–dc23
    LC record available at https://lccn.loc.gov/2018046543

ISBN 978-1-108-42201-7 Hardback
ISBN 978-1-108-43447-8 Paperback

Contents

Preface

Islands have always been at the forefront of evolutionary thinking. They played a pivotal role in the initial development of the theory of evolution via natural selection and have continued to teach us new lessons about evolution ever since. The aim of this book is to accelerate interest in what I believe to be a neglected aspect of island biology, the island syndrome in plants. It searches for repeated patterns in plant form and function on islands using the hypothetico-deductive research method in an effort to guide the field away from just-so natural history storytelling and help transform it into a more rigorous, scientific discipline.

I first became interested in the island syndrome in plants as an undergraduate student at the University of California, Berkeley, 25 years ago. I was enrolled on a course on biogeography and the fourth edition of Cox & Moore's *Biogeography: An Ecological and Evolutionary Approach* was assigned as required reading. On page 122, I discovered a series of line drawings illustrating six tree species that were endemic to St Helena Island, in the South Atlantic Ocean. All were shown to produce composite flowers, a hallmark of many common weeds. However, these 'weeds' had strikingly stout trunks and a distinctive, sturdy look about them.

Remarkably, Cox & Moore explained that all six species evolved from weeds on the mainland. This sent my mind wondering. What was it about this tiny speck of land in the middle of the ocean that turned them into giants? I suddenly wanted to visit St Helena to have a look for myself. Cox & Moore then went on to explain that this 'weeds-to-trees' evolutionary pathway was actually quite common. Similar-looking trees had evolved on many isolated islands across the globe. It was insularity in general, rather than some idiosyncratic feature of St Helena, that turned these weeds into trees.

I suppose my mind has wondered about the principles underpinning Cox & Moore's illustration ever since. In the years that followed, I somehow managed to make a career out of trying to solve the evolutionary riddles posed by island plants. My greatest hope for this book is to inspire future students to wonder about the island syndrome in plants, as Cox & Moore inspired me.

This book is split into seven chapters. Five of these chapters focus on a specific aspect of the life history of plants, bracketed by an introduction and conclusion. Some of these chapters differ widely in length, for good reason. For example, Chapter 2 ('Differences in Defence') is particularly long because most previous research on the subject has focused on single archipelagos, and is, therefore, disjointed and in need of an extensive overview. On the other hand, the topic of Chapter 4 ('Gender & Outcrossing') has been studied rigorously for decades and reviewed on multiple occasions. So, instead of another review of the topic, it provides a brief summary that focuses on several underappreciated aspects of the reproductive biology of island plants using new data. Chapter 6 ('Loss of Fire-Adapted Traits') is also comparatively short, but for a different reason again. Very few studies have investigated the subject quantitatively, so there is less to cover.

Any book on the island syndrome presents a serious challenge to just one author. Although the topic focuses on a single aspect of the earth's geography, the topic is biologically very broad, encompassing a wide range of subjects, from defence and dispersal to reproduction and morphology. I have done my best to cover these varied aspects of biology both thoroughly and accurately, but in those instances where I have failed, I apologise to experts in each of these subjects, whose research I neglected to discover, appreciate, and discuss accurately.

I am indebted to many people for their help. Thanks most of all to my family for their support and encouragement. My colleagues have been invaluable, particularly here at Victoria University of Wellington. After moving to New Zealand, I was very lucky to find myself surrounded by a group of enthusiastic emeritus advisors, including George Gibbs, Phil Garnock-Jones, and Ben Bell, who patiently answered all of my questions about the natural history of New Zealand. John Dawson was particularly helpful in this regard. He also encouraged me to expand my research horizons to include New Caledonia, the Chatham Islands, and islands further afield. Murray Williams has also been endlessly supportive, painstakingly reviewing all seven chapters of this book. Ian Hutton has been similarly helpful on Lord Howe Island. Cherie Ball, Matt Biddick, and Bart Cox provided unpublished data. Pictures were kindly provided by Brent Alloway, Phil Garnock-Jones, Ian Hutton, Edward Farmer, and Philip Thomas. Robert Cross helped with the illustrations, and Karson and Luke helped with referencing. Chris Fink helped with Box 1.1, and Matt Ryan helped with Box 1.2. Brent Alloway (Chapter 1), Bill Lee (Chapter 2), John Lambrinos (Chapter 3), Janice Lord (Chapter 4), Mark Lomolino (Chapter 5), and Phil Rundel (Chapter 6) provided valuable comments to individual chapters. However, all errors, mistakes, and omissions are entirely my own.

K C Burns
Wellington, New Zealand

1   Introduction

Emblematic Island Animals

This chapter provides an introduction to some general concepts in biogeography and evolution that are needed in order to digest the rest of the book. However, unlike subsequent chapters, it will focus on animals rather than plants. Three island animals in particular, when considered together, illustrate a variety of evolutionary trends that appear to comprise an 'island syndrome'. The first emblematic island animal is a large flightless rail from New Zealand known as takahe (*Porphyrio hochstetteri*: Rallidae), which is perhaps the best living example of insular naivety and flightlessness. When compared with other flightless island birds, takahe also illustrate the subtle, yet important, differences between parallel, convergent, and repeated pathways of evolution. Giant tortoises (Testudines: Testudinidae) come next. They illustrate how gender and outcrossing might vary during the process of island colonisation, and how species interactions can shape the course of island evolution. Giant tortoises also illustrate the difference between anagenesis and cladogenesis, and, perhaps most importantly, they are a poignant reminder of how scientific progress hinges on an objective and rigorous research philosophy. Lastly, from our own evolutionary lineage, is the controversial 'hobbit' from Flores Island in Indonesia (*Homo floresiensis*: Hominidae), whose diminutive stature has been a source of intense scientific scrutiny and debate. When viewed jointly, these three emblematic island animals illustrate many of the trends in plant life history that are explored in subsequent chapters. They also hint at the possibility that plants and animals might travel down the same evolutionary pathways on isolated islands.

TAKAHE

Takahe are far from ordinary birds. They look like feathered circus clowns, with purple, pear-shaped bodies, big red bills, and floppy feet that keep them firmly on the ground (Fig. 1.1). Takahe are also strikingly ambivalent to onlookers – as wild animals, they show little regard for their own safety when surrounded by people. However,

(a)

(b)   (c)

FIGURE 1.1 (a) Takahe (*Porphyrio hochstetteri*: Rallidae), (b) dodo (*Raphus cucullatus*: Columbidae; photo taken by Dorling Kindersley/Getty Images), and (c) kakapo (*Strigops habroptilus*: Strigopidae; photo taken by Robin Bush/Getty Images), which together provide a striking example of repeated evolution of predator naivety, flightlessness, and insular gigantism.

the takahe's position in the world of science is far from funny. Not only have they generated one of the 'greatest moments in ornithological history' (Fitzpatrick 2001), the evolutionary lineage to which they belong provides what is arguably the best example of repeated evolution in animals.

Differences in Defence

By the turn of the twentieth century, the takahe (*P. hochstetteri*) was thought to be extinct. In 1898, the 'last' bird was caught in a mountainous region of New Zealand's South Island by a hunter's dog. What was left of this unfortunate bird was then sold to a museum in Europe, as was common practice at the time (Balance 2001). Many New Zealand bird species were being driven rapidly to extinction at that time by changes brought by humans, and their rarity made them valuable to collectors.

Many other New Zealand bird species were in a similar predicament to the takahe. Since the arrival of people less than 800 years ago, approximately half of the archipelago's avifauna has gone extinct (Tennyson & Martinson 2006). Many factors have contributed to the demise of New Zealand's avifauna, but two factors stand out from the rest. Early accounts labelled many native birds, and the takahe in particular, as 'delicious' (Balance 2001). Many ended up on the dinner table. But perhaps even more damaging was the introduction of other mammalian predators. Rats, cats, and weasels (and many others) regularly accompany us on our travels to isolated islands and subsequently eat their way through the local checklist of island animals (Wilson 2004). A wave of extinction has coincided with the arrival of humans to virtually every archipelago on the planet, from Hawai'i to Mauritius, due in large part to the introduction of mainland predators (Steadman 2006; Duncan et al. 2013).

So why are island animals so susceptible to mainland predators? One factor seems obvious. Large stretches of open ocean present a formidable challenge to organisms colonising islands on their own, and only certain types of animals pass the test, dispersing to islands and becoming residents. While birds and reptiles are good at dispersing over the open ocean, mammals are not. With the exception of bats, terrestrial mammals are woefully equipped for traveling long distances over the open ocean, mainly because they cannot survive long periods without food, a prerequisite for long-distance dispersal on ocean currents. As a result, animals that are better at colonising islands walk an evolutionary pathway without the company of mammalian predators and they are wholly unprepared for the predatory onslaught when they arrive (Box 1.1).

In a way, the takahe was lucky. Half a century after the 'last' bird was sold to science, an amateur naturalist named Geoff Orbell began to comb the backcountry in New Zealand's Southern Alps in the hopes of finding one alive. In April 1948, his efforts paid off, much to the delight of bird lovers around the globe, some of whom labelled the discovery as the 'greatest moment' in the history of ornithology (Balance 2001; Grueber & Jamieson 2011).

Conservationists have learned a lot from the takahe's plight. The vulnerability of island biotas to mammalian predators is now widely recognised, and what's left of New Zealand's natural history is now painstakingly protected from mammalian predators. Intensively managed, mammal-free nature reserves are the only places that the takahe are now capable of calling home (Jamieson & Ryan 2001; Lee 2001).

Although the absence of adaptations to thwart mammalian predators characterises many island animals (Blumstein 2002), this does not mean that island animals are completely defenceless. Vertebrate predators do reach isolated islands, but they often have feathers instead of fur. Ironically, the biggest raptor known to science, Haast's eagle (*Harpagornis moorei*), evolved in New Zealand. Correspondingly, New Zealand birds often possess traits that help them avoid being eaten by bird predators rather than mammalian predators.

Birds hunt differently from mammals, relying more on sight than smell. Therefore, a good way to avoid avian predators is to evolve colouration that mirrors your surroundings, rather than concealing

BOX 1.1 Dispersal disharmony

Dispersal to isolated islands is governed primarily by chance. However, chance dispersal is more likely in some types of organisms than it is in others. The likelihood of dispersal to an isolated island is affected by two factors: over-water dispersibility and survivability during transit. Taxa with high dispersibility and/or high survivability are much more likely to populate isolated islands relative to taxa with low dispersibility and survivability, provided islands have suitable habitat for successful establishment (see Carvajal-Endara et al. 2017).

Dispersibility refers to an organism's rate of travel. Species that are capable of flight, most notably birds, have excellent dispersibility. Spiders which 'balloon' to new localities by spinning long dispersal threads that carry them passively on air currents also have high dispersibility. At the other end of the spectrum are reptiles and mammals, both of which cannot fly (aside from bats, which are functionally similar to birds in this regard). They have to either swim or float to islands – a much slower mode of transport.

Taxon-specific dispersibility can be described by a function $D(t)$, which characterises the distance travelled, D, through time, t (Fig. B1.1a). Assuming a constant rate of travel, $D(t)$ can be defined as a straight line with an intercept of zero and slope, ϕ, which varies among taxa according to their rate of travel, $D(t) = \phi t$. Birds have a higher value of ϕ than both non-volant mammals and reptiles, which are both

(a)

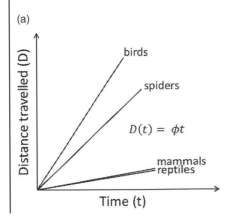

FIGURE B1.1A Relationships between the distance dispersing propagules travel per unit time in birds, spiders, mammals, and reptiles.

BOX 1.1 **(cont.)**

at the mercy of ocean currents and are therefore represented by $D(t)$ curves with shallower slopes. Spiders have an intermediate value of ϕ, given that they can fly, albeit passively, so they are at the mercy of air currents.

Survivability refers to an organism's mortality rate during transect. It can be described by a second function. $S(t)$, which characterises the number of surviving propagules, S, per unit time, t (Fig. B1.1b). $S(t)$ can be defined as a straight line with an intercept, P, which reflects the number of propagules present on the mainland that initiate the process of dispersal (i.e., *propagule pressure*), and a slope, α, which represents the mortality rate during dispersal, $S(t) = P - \alpha t$. Spiders should have a higher value of P than do reptiles, birds, and mammals, because they are present in higher densities across most continental landscapes. Reptiles should have relatively large (i.e., less negative) values of α, while birds and mammals should have relatively small (i.e., more negative) values of α because reptiles are ectothermic and can survive for longer periods without food or water relative to endothermic birds and mammals. Spiders might be expected to have intermediate values of α; although they are ectothermic, they have higher metabolic rates, shorter generation times and higher mortality rates, given they are smaller in size than reptiles.

Considering taxon-specific dispersibility and survivability jointly, the probability of chance dispersal is determined by three factors,

(b)

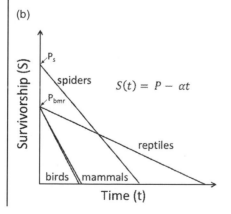

FIGURE B1.1B Relationships between the number of dispersing propagules surviving through time in birds, spiders, mammals, and reptiles.

$S(t) = P - \alpha t$

BOX 1.1 **(cont.)**

(c)

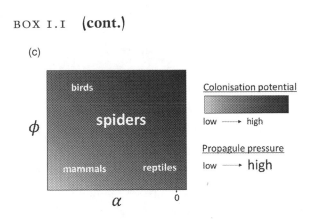

FIGURE B1.1C Differences in disperibility (ϕ) and survivability (α) in birds, spiders, mammals, and reptiles. Darker regions refer to greater colonisation potential and larger font size refers to higher propagule pressure (P).

ϕ, P, and α. It increases with ϕ and P, how fast a species can move across the landscape and how many individuals embark on the journey, respectively. It also increases with α, the taxon's probability of mortality during transit. These three parameters for birds, mammals, reptiles, and spiders can be potted together in a single figure representing the probability of chance island colonisation by each taxon (Fig. B1.1c).

olfactory cues. Although their bills are red and their flanks are purple, the top of the takahe's back is covered in feathers that are a greenish, mottled-brown colour, which may have rendered the takahe less conspicuous to predatory raptors trying to spot them against a background of greenish-brown vegetation from above. This is unusual in the evolutionary lineage to which the takahe belongs. Most of the takahe's continental cousins produce only blue and black feathers.

Mainland birds that evolved alongside mammalian predators often emit very little odour, giving predators that hunt by smell little to work with. On the other hand, freed from selection by predators that rely mostly on their sense of smell, many New Zealand birds 'stink' a bit, which has the unfortunate consequence of making them easier for introduced mammalian predators to detect (Azzani 2015).

As we shall see in Chapter 2, plants and takahe may not be so different when it comes to defending themselves against being eaten. Plant adaptations to thwart mammalian predators, such as thorns, are hard to find on many islands, perhaps echoing the takahe's ambivalence to humans and other predatory mammals. On the other hand, island plants often display a suite of distinctive traits that are rare or absent on continents, which may have helped them avoid being eaten by unique island herbivores, most of which are now extinct. Given that many island herbivores are no longer with us, research into plant defences on isolated islands has a 'whiff of mystery' (Zotz et al. 2011) and has sparked decades of scientific debate and controversy.

Loss of Dispersibility

The takahe's closest living relatives are found in open, marshy habitats throughout Australia, Southeast Asia, and parts of Europe and Africa. Unlike its comical island cousin, the purple swamphen or 'pukeko' (Porphyrio porphyrio) is always wary of onlookers and keeps its distance from would-be predators. The pukeko and takahe also differ morphologically. The pukeko's bill is more modest than the takahe's and is better suited to its omnivorous diet, which includes grasses, seeds, and small animals. The takahe's bill is much larger and more heavily reinforced to suit a more specialised vegetarian diet (Mills & Mark 1977). In the summer, takahe forage above the tree line in alpine environments for their favourite food: the juicy tillers of tussock grasses. After pulling grass tillers out of the ground with their bill, they transfer them to their talons and gnaw on their nutrient-rich stalks, a bit like a parrot eating seeds from its feet. It takes quite a bit of effort to unearth the tillers of alpine tussocks (Chionochloa spp.). So a stoutly reinforced bill seems to have evolved to help it overcome its hard-to-break prey. A big, stout bill also helps the takahe forage efficiently in winter, when they travel downslope, below the treeline.

Here they dig up fern rhizomes, which they consume using the same, parrot-like technique.

Their feet are different too. The pukeko has long, spindly toes, which are ideal for walking along the top of soft, muddy surfaces in marshlands. The takahe's talons are short and stout, which better enable them to manoeuvre around tussocks and stunted shrubs on hard, rocky ground. Stout feet and legs are also required to support heavier bodies, and they are less susceptible to freezing temperatures.

Differences in bill and talon morphology between the takahe and pukeko are interpretable in light of differences in their preferred habitats, but these differences pale in comparison to differences in their body mass and wing length. The takahe are two to three times heavier than the pukeko, yet their wings are smaller. A bird's ability to fly is determined in large part by its 'wing-loading ratio', or the ratio of wing size to body mass (Van den Hout et al. 2010). So, as the size of a bird's body increases, the size of its wings must keep pace in order for it to fly. But, in the case of the takahe, disproportionate changes in body size and wing length have channelled it down an evolutionary trajectory where it can no longer take to the air.

At first glance, the presence of flightless animals on isolated islands seems strange. Flight is arguably the best way for organisms to disperse across inhospitable terrain. So how did the takahe come to call New Zealand home? Molecular analyses show that the takahe diverged from the pukeko long after New Zealand split from Australia geologically (Trewick 1997; Garcia-R & Trewick 2015). So, a common ancestor to both the pukeko and takahe probably flew to New Zealand from Australia and subsequently lost the ability to fly. But, given its obvious adaptive advantages, why would evolution then favour the loss of flight on islands, or anywhere else for that matter?

A likely explanation for insular flightlessness is that the selective advantages of functional wings are relaxed in island environments (Lahti et al. 2009). Ground-dwelling predators were often rare or absent on islands (Livezey 2003). So, once freed from selection by ground-dwelling predators, the surplus energy that was previously devoted to wing development can instead be 'invested' elsewhere morphologically (McNab & Ellis 2006).

Another possible explanation to the loss of flight in the takahe relates to the costs associated with being lost at sea. Islands are very windy places (Box 1.2). In the absence of terrestrial topography to slow wind speeds down, wind whips across the ocean surface at greater speeds than on continents. Consequently, flightedness could be directly disadvantageous on islands, given their close proximity to open ocean, if big wings make island animals more susceptible to being accidentally lost at sea (Darwin 1859; Cody & McC. Overton 1996). This hypothesis will be the focus of chapter 3, which reviews the evidence for the loss of dispersibility in plants.

Alternatively, insular flightlessness could also relate to the stability of island environments. Birds inhabiting continents often track changes in the landscape over large spatial scales (e.g., Burns 2002; 2004). For example, in fire-prone ecosystems, birds can fly over large distances not only to avoid the flames, but to also to recolonise sites as resource availability increases during post-fire succession (Loyn 1997; Watson et al. 2012). However, space is limited on islands and wildfires are less common. Therefore, increased dispersibility to avoid fire and track post-fire changes in resource availability are less advantageous on islands than on the mainland.

Sedentary organisms, and plants in particular, must cope with fire differently. Rather than avoiding fire, plants in fire-prone ecosystems display a variety of traits that help them either resist fire damage or be resilient to wildfires, given that they cannot flee from its harmful effects (Box 1.3). Whether plants repeatedly lose fire-adapted traits after colonising isolated islands will be explored in Chapter 6.

Parallel, Convergent, & Repeated Evolution

Subfossil bones of takahe-like birds have been found on both the North and South Islands of New Zealand. This led to initial speculation that, prior to the arrival of humans, a single species of takahe occurred throughout New Zealand. If so, it would suggest all takahe

BOX 1.2 Island environments

Islands, by definition, are surrounded by water, so the chemical properties of water have a greater impact on island environments than on continents. H_2O is the only chemical on earth that naturally exists in all three states (i.e., solid, liquid, and gas). When water vapour is converted into water, or when water is converted into ice, energy is released. Conversely, energy is required to convert ice into water, and water into water vapour. The energy involved in the conversion from one state into another is known as *latent heat*, and the latent heat of H_2O is quite high. As a result, the oceans absorb, and store, huge quantities of the sun's energy. This has two important effects on island environments. It both lowers overall temperatures and has a dampening effect on temperature fluctuations. Therefore, islands have slightly cooler, yet more stable, temperature regimes than continents (Fig. B1.2).

The world's oceans are also the birth place of precipitation. With the input of energy from the sun, water is transformed into water vapour, which rises into the atmosphere. Therefore, the amount of water vapour in the atmosphere above the ocean tends to be relatively high

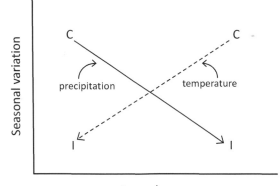

FIGURE B1.2 Differences in temperature (dashed line) and precipitation (solid line) between islands (I) and continents (C). Seasonal variability is shown on the y-axis and annual means are shown on the x-axis. Islands tend to be cooler and wetter than continents and fluctuations in temperature and precipitation are also less pronounced. See Weigelt et al. (2013) for precise values and global statistics.

BOX 1.2 **(cont.)**

and declines towards the centre of continents. When water vapour is forced upwards, it loses energy (i.e., it cools) and falls back to the earth as precipitation. So, when air is forced aloft by mountains, precipitation increases, a process known as orographic rainfall. Mountainous areas close to the ocean typically experience high precipitation, and some of the wettest places on Earth are on high-elevation islands. Islands also experience more stable levels of precipitation throughout the year (see Weigelt et al. 2013).

A third difference between the climate of islands and continents results from differences in topography. Because oceans are liquid, gravity pulls them into flat surfaces. On the other hand, a variety of geologic processes (e.g., plate tectonics, volcanism) create variation in relief in the earth's solid surfaces, ranging from low-lying valleys to tall mountains. As a result, wind generally encounters greater friction with the earth's surface on continents. With little topographic relief to slow winds down, wind speeds are much higher over the ocean (Archer & Jacobson 2005). In addition to differences in surface friction between oceans and continents, winds are generated at sea from energy fluxes occurring between the ocean and atmosphere. As a result, islands are much windier places than continents.

evolved from a single, pukeko-like ancestor, and large bodies, small wings, big bills, and stouter legs would have evolved just once. However, more recent molecular analyses have shattered this rather tidy explanation.

Despite their morphological similarity, the North and South Island takahe are actually only distantly related. Their closest ancestors are actually the flighted pukeko, rather than each other (Trewick 1997). This suggests that, at different points in the evolutionary past, pukeko-like birds arrived independently on New Zealand's two main islands. Each population then travelled along separate, yet analogous, evolutionary pathways. Each lost its fear of mainland predators,

BOX 1.3   **Fire-adapted traits**

Wildfires are a common phenomenon across most of the planet. All biomes experience fire at least occasionally (Chuvieco et al. 2008; Bowman et al. 2009). As a result, it's not surprising that both plants and animals have evolved various ways to cope with wildfires. Very little is known about how animals have evolved in responce to fire (Pausas & Parr 2018). However, given their mobility, many animals can escape the immediate effects of fire by flying or running to safety. Plants are not so lucky. Because they are sedentary, they tolerate wildfires either by being resistant to injury caused by fire or resilient to fire damage.

Plants display a diverse suite of traits that can help them cope with fire, many of which lack obvious parallels in animals (Keeley et al. 2011). For example, some plants have evolved thick bark that protects their tissues against heat damage (Pausas 2015). Others produce insulating fruits or cones to protect their seeds (i.e., serotiny) (Causley et al. 2016, Fig. B1.3). Plants can also re-sprout after fire, either from lateral meristems or from specialised bud banks below ground (i.e., lignotubers) (Pausas & Keeley 2017; Pausas et al. 2018).

FIGURE B1.3 Examples of serotinous fruits in the genus *Hakea*, Proteaceae from eastern Australia (*Hakea sericea*, left; *Hakea propinqua*, second from left; *Hakea dactyloides*, second from right; *Hakea teretifolia*, right). Despite marked differences in gross morphology, the fruits of all four species usually remain closed on parent plants for long periods, and open to release two winged seeds following fire-induced mortality.

BOX 1.3   **(cont.)**

Although serotiny, thick bark, below-ground bud banks, and fire-resistant seeds are known to help plants cope with fire, it is often difficult to establish whether fire has selected for them directly because they can potentially serve other functions. For example, thick bark can insulate plants against heat damage (Pausas 2015) as well as cope with a range of other stressors, including pathogenic attack or storm damage. As a result, establishing direct links between fire and faire-adapted traits can be difficult (Keeley et al. 2011; Rosell 2016).

Isolated islands are among the least fire-prone areas on earth, for several reasons. Islands tend to receive greater quantities of rainfall than comparable sites on the continents, and rainfall tends to fall more regularly (see Box 1.2). Islands are also cooler and annual temperature fluctuations are less pronounced. Islands are also less connected to surrounding areas by fire, given that lightning strikes at sea will not propagate fire to nearby islands. As a result, we might expect island plants to show weaker evidence for adaptations to tolerate wildfires, given that they tend to occur less frequently on islands.

evolved bigger bills, stouter feet, heavier bodies, and useless wings all on its own. None of these unusual traits result from shared ancestry.

Remarkable as this may seem, this is not the end of the repeated nature of the takahe's evolutionary history. Seven hundred kilometres off the east coast of Australia lies Lord Howe Island, a tiny speck of land that was not discovered by humans until it was stumbled upon by a supply ship that departed from Sydney in 1788. As humans stepped ashore for the first time, they were greeted naively by a takahe-like species of *Porphyrio* that shared many of the characteristics of the two lineages of takahe in New Zealand. Compared with the pukeko, its body was a bit bigger, its feet somewhat stouter, its bill bigger, and its wings shorter (van Grouw & Hume 2016). However, the Lord Howe Island form (*Porphyrio albus*) was more similar to the pukeko than both takahe in New Zealand, suggesting it had walked their shared evolutionary pathway for less time than the two New

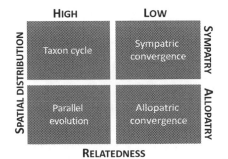

FIGURE 1.2 Types of repeated patterns in insular evolution. The left column refers to closely related species, while the right column refers to distantly related species. The top row refers to species that are distributed sympatrically (together), while the bottom row refers to species that are distributed allopatrically (apart). The taxon cycle refers to repeated patterns in the evolution of related species that occur on the same island, but arrive at different points in time. Sympatric and allopatric convergence refer to repeated patterns in the evolution of distantly related species, on the same and separate islands, respectively. Parallel evolution refers to repeated evolution in related species on different archipelagos.

Zealand forms. Sadly, it was not long before sailors ate the last one and we are left with just two stuffed museum specimens. More remarkably still, this same story of repeated evolution in pukeko-like birds seems to have occurred on several other islands, both in the Pacific and Indian Oceans, although the scientific details are more sparce (Slikas et al. 2002; Livezey 2003; Kirchman 2012).

Repeated patterns in evolution can be classified into distinct categories (Fig. 1.2). Evolutionary trends in the genus *Porphyrio* provide a fine example of *parallel evolution* – when related species evolve similar characteristics in different geographic locations. A subtly different category of a repeated evolution is termed *convergent evolution* – when distantly related organisms travel down the same evolutionary pathway to arrive at the same place morphologically.

The dodo (*Raphus cucullatus*: Columbidae) is one of the most recognisable island animals (Cheke & Hume 2008). Standing one metre above the ground, it was huge by bird standards, far too big to fly given its stumpy little wings (Fig. 1.1). It had a massive bill, which it used to dig in the ground for roots, rhizomes, and bulbs. Its name is also thought to come from the Portuguese word for 'fool' because it failed to recognize the danger posed by early European sailors who came ashore in its native Mauritius, famished after a long journey around Cape Horn into the Indian Ocean. It shares no relation to the takahe phylogenetically, aside from the fact they are both birds (Shapiro et al. 2002). The dodo evolved from a flighted, fruit-eating, forest-inhabiting pigeon, making its relationship with the takahe an example of convergent, rather than parallel evolution.

Although the takahe and dodo exemplify convergent evolution, textbook definitions of repeated patterns in evolution, whether they be parallel or convergent, leave a lot to be desired. Both parallel and convergent evolution refer to the same conceptual circumstance – the evolution of phenotypic similarity via shared selection pressures. What delineates the two is the degree of ancestral similarity. However, the degree of ancestral similarity required to tip the balance from parallel to convergent evolution is arbitrary and sometimes difficult to resolve.

A third example of convergent evolution highlights another problem with traditional terminology. Like the dodo and the takahe, the kakapo (*Strigops habroptilus*: Psittaciformes) is another large, naïve, and flightless island bird (Fig. 1.1). Kakapo almost went extinct after humans reached its native New Zealand, yet it is very well prepared for avian predators. It has mottled-green plumage that is similar in appearance to herbs and leaf litter on the forest floor. It freezes when threatened and has a sweet, musty odour (Hagelin 2004). While this probably was not a problem prehistorically, it's now a beacon to introduced mammalian predators that hunt using olfactory cues. Quite remarkably, however, the kakapo is a parrot, rather than a rail or pigeon. It would, therefore, appear to fall under the category of convergent, rather than parallel evolution. But the takahe and kakapo evolved to be convergent in the same place (New Zealand), while the takahe and dodo evolved in different places (New Zealand and Mauritius, respectively), a distinction that is not incorporated into traditional definitions of convergent evolution.

Integrating this difference into subdivisions of repeated evolutionary patterns (Fig 1.2), the takahe and dodo provide an example of *allopatric convergence*, or the evolution of similar morphology in distantly related organisms in different geographic locations. On the other hand, the takahe and kakapo provide an example of *sympatric convergence*, or the evolution of similar morphology in distantly related organisms in the same place. Importantly, the underpinning evolutionary circumstances are the same in both instances.

Traditionally, it has always been assumed that allopatry is a key requirement for parallel evolution to take place. This makes sense. If related taxa arrive in a new location at roughly the same time, before barriers to gene flow had evolved, they would interbreed and evolve in unison as the same species. However, this need not necessarily be the case (Savolainen et al. 2006).

Imagine a plant species that colonises an island after surviving a long journey overseas. Let's assume that it's a thistle, which are herbaceous plants that typically have spiny leaves and wind-dispersed seeds. After arrival on the island, it begins to lose its sharp leaf spines, which are no longer necessary in the absence of large herbivores. Its fruit morphology also evolves to better suit its new island home. The feathery plumes that carry its seeds into the air evolve to become smaller, such that the seeds lose their capacity to disperse very far. If the thistle travels this island evolutionary pathway for long enough, it will become not just morphologically isolated from the mainland population, but genetically isolated as well. At this point, the process can start again, if an individual from the original, mainland population is lucky enough to make the long journey out to the same island once again. Afterwards, it would be subjected to the same selection pressures, following in the footsteps of its earlier colonising sister species down the same evolutionary pathway, leading to parallel evolution on the same island.

This process, the sequential colonisation of an island by similar mainland taxa at different points in time, has traditionally been embedded within its own theoretical framework, which E O Wilson

(1961) termed the *taxon cycle*. In his travels throughout the Pacific, Wilson noticed that ants (Formicidae: Hymenoptera) inhabiting lowland habitats were often similar to taxa inhabiting analogous habitats in Australia, New Guinea, and Southeast Asia. However, as he walked inland and up the slopes of high islands, he usually encountered taxonomically distinct species with reduced ranges. From this he deduced a cyclic pattern in island colonisation and evolution. Ants appeared to be colonising islands from the same ancestral populations in maniland Australasia. Afterwards, they expanded their range and evolved to use inland habitats, often speciating in the process, thus establishing a cyclic pattern of evolutionary change.

Although he was primarily interested in distributional patterns, Wilson (1961) also noticed that island ants tended to change both morphologically and behaviourally in apparently predictable ways as they shifted their use of island habitats. They tended to lose spinescent structures and colony sizes of eusocial species tended to decline. In other words, they tended to lose defensive adaptations and evolve different 'body' sizes. Subsequent studies of taxon cycles have focused on species distributions rather than their morphology (Ricklefs & Cox 1978; Maclean & Holt 1979; Ricklefs & Bermingham 2002; Jønsson et al. 2014; Economo et al. 2015). So, the evolutionary processes encapsulated by the taxon cycle are generally considered outside the context of parallel or convergent evolution. Yet, the three phenomena are clearly related conceptually (Fig. 1.2).

While it's useful to understand the distinguishing features of parallel evolution, convergent evolution, and the taxon cycle, for much of this book it will not be needed. All of them characterise the same basic principle – repeated patterns in evolution. The goal of this book is to explore these patterns in plants, regardless of their classification. This goal is unusual in today's scientific world. Most work in ecology and evolution seeks to understand the origins and maintenance of biological diversity. We will explore the other side of this scientific coin – biological regularity in the form of repeated evolution.

GIANT TORTOISES

At first glance, giant tortoises and flightless rails could not be more different. Tortoises do not grow feathers, birds do not have shells. Tortoises are ectothermic, birds are endothermic. Tortoises are world famous for being slow, while the fastest animal on the planet is a bird (the peregrine falcon, *Falco peregrinus*). These are just a few of the things that distinguish tortoises from birds. However, when you look beyond the traits that confine them to their respective evolutionary lineages, giant tortoises and takahe exhibit some remarkable similarities – similarities that suggest insular environments have shaped them in analogous ways.

Just as takahe are bigger than pukeko, giant tortoises are much larger than their closest relatives on the mainland. Reminiscent of the takahe's big bill and stumpy feet, shell shape often differs markedly among giant tortoises – differences that help them exploit particular island habitats. Tortoises are also fantastic island colonists, but for very different reasons than birds. Tortoises are not great swimmers, but they float effortlessly, and their lethargic nature is key to their success in overseas travel. Given their sluggish metabolic rate, they can survive long periods without food or water, which is an important asset in long-distance travel by floating on the surface of the sea. So sea-swept tortoises have floated at the mercy of ocean currents in all of the world's major oceans, until they were eventually lucky enough to cross the path of an isolated archipelago and were washed ashore.

Their highly successful, lethargic propensity to travel the world's oceans, coupled with a remarkable degree of morphological similarity between distantly separated tortoises, led early naturalists to surmise that all giant tortoises evolved from a single common ancestor and then island hopped across the globe. However, this scenario did not stand up to closer scientific scrutiny. Just as flightless rails evolved in parallel in different archipelagos, giant tortoises in the Atlantic, Caribbean, Indian, and Pacific Oceans may look similar, but are distantly related.

Another attribute linking giant tortoises and flightless rails is they taste good. Early island explorers coveted giant tortoises as food, not just because they were 'the most delicate morsel one could ever eat' (Dubois 1674, cited in Cheke & Hume 2008), but also for the same reason tortoises succeeded in colonising isolated islands in the first place – they could survive long periods without food or water. Early sailors took advantage of their slow metabolism and packed giant tortoises into the hulls of their wooden sailing ships in their hundreds, where they clung to life for months, providing fresh meat for hungry crews as it was needed. So, while their tenacity for life when deprived of food helped tortoises populate the world's islands passively, it played a big role in their demise.

Luckily, several species of giant tortoises escaped extinction on islands, in both the Indian (Seychelles) and Pacific (Galapagos) Oceans. Yet, both archipelagos previously housed many more species of tortoise than can be found there today. Giant tortoises also occurred in the Canary, Caribbean, and Mascarene Islands.

Gender & Outcrossing

Even though tortoises can drift on the open ocean for months at a time, they need to be very lucky indeed to randomly intercept an isolated island. But, upon closer inspection, a tortoise that was fortunate enough to be washed ashore on an isolated island would not be so lucky after all. If the island had yet to be colonised by this species, any new arrival would be all alone. Without a mate, the island would revert back to being 'tortoise-free' after the lonely traveller died, as the chance that a tortoise of the opposite sex would make the same remarkable journey at the same time is very poor indeed. Nevertheless, some lonely island colonists have a trick up their sleeve to circumvent the necessity of a mate – they blur the distinction between the sexes.

Most populations of vertebrates are comprised of individuals that are either strictly male or strictly female. Individuals from these *dioecious* species require genetic material from both a male and a female parent. In order for a dioecious species to found a new, self-sustaining population, at least one male and one female have to immigrate together (unless it is founded by a gravid female who mated with a male on the mainland prior to island colonisation). Although useful in avoiding inbreeding depression (see Barrett & Hough 2012), dioecy is not very helpful when dispersing to isolated islands. Under these circumstances, it might pay to be hermaphroditic.

Hermaphrodites are capable of both male and female function. Freed from the need of another individual, they are at an obvious advantage when colonising isolated islands. So hermaphroditism may often be associated with lineages of organisms that populate isolated islands.

Interestingly, a closer look at the reproductive biology of giant tortoises shows that they can sometimes be hermaphroditic. Kuchling & Griffiths (2012) inspected the gonads of 106 giant tortoises inhabiting Aldabra Island (*Aldabrachelys gigantea*), and they found that three animals had both testes and ovaries. While the overall number of hermaphrodites is low at present, hermaphroditism could have potentially played a key role in their success in immigrating to Aldabra Island at some point in the past.

Populations of many types of animals vary in their degree of gender dimorphism, and selection may favour different genders under different ecological circumstances (Ghiselin 1969). For example, we might expect for hermaphroditism to be favoured in the initial stages of island colonisation, when mates are in limited supply. However, immediately following island colonisation and population expansion, we might expect this situation to reverse (see Baker 1955, 1967; Charnov 1979; Puurtinen & Kaitala 2002). As the newly founded population grows, and mate availability increases, dioecy may be favoured to promote outcrossing.

Unlike animals, plants are more commonly hermaphroditic. Over 70% of all plant species are comprised of individuals that produce both male and female reproductive structures. However, hermaphroditic plants have evolved a rich diversity of mechanisms to prevent self-pollination and promote outcrossing. The reproductive biology of island plants is explored in Chapter 4, which will attempt to establish general trends in floral biology and outcrossing.

Species Interactions

When Darwin first arrived in the Galapagos, he was greeted by Nicolas Lawson, the vice governor of the islands. Unlike Darwin, Lawson had been in the Galapagos for long enough to carefully inspect the shells of tortoises living in different parts of the archipelago. After months of careful scrutiny, he was able to discern distinctive differences in the shape of shells produced by tortoises living in different environments, which led him to boast that if someone brought him a tortoise, he could identify from where it came in the archipelago (Nicholls 2014).

Generally speaking, giant tortoises produce shells that are shaped in one of two ways. The first type of shell shape slopes steadily downwards at the front, sides and back, resulting in a uniformly rounded appearance. The second arches dramatically upwards at the front, away from the animal's head and neck, allowing it to stretch its head upwards (Fig. 1.3).

FIGURE 1.3 'Lonesome George', the last surviving Pinta Island giant tortoise (*Chelonoidis abingdonii*) from the Galapagos Islands. (photo from Getty Images)

Recent research suggests that the shape of tortoise shells may help them self-right after they topple over (Chiari et al. 2017). However, giant tortoises are almost exclusively herbivorous and the shape of their shells is also related to the spatial distribution of their food plants. Tortoises with saddle-shaped shells are typically found in drier sites that are dominated by scrubby trees and shrubs, where food commonly comes in the form of leaves, fruits, and flowers located above the ground. Freed from the restrictions of a downward-arching shell, some saddle-shaped tortoises in the Galapagos can reach a whopping two metres above the ground when they are hungry (Nicholls 2014). On the other hand, domed-shaped tortoises commonly occur in areas with higher precipitation and denser populations of herbaceous plants. Given the small stature of herbaceous plants, tortoises in this type of habitat do not have to reach upwards to find their next meal, and the overhanging shape of their shells is not a limiting factor in finding food.

On the other side of this evolutionary coin, plants seem to have evolved characteristics that help them avoid being eaten by giant tortoises. The pad-like stems of 'prickly-pear' cacti (*Opuntia* spp.) are a particular favourite of giant tortoises in the Galapagos. Their leaves, which have been shaped by evolution to be sharp, rigid spines, have been selected to deter herbivores. Strictly speaking, when a tortoise consumes a leaf, it comes under the heading of parasitism – the tortoise benefits and the plant is harmed, but not killed. If the tortoise consumes enough plant tissue that it results in the death of the plant, then it would be reclassified as predation. Often times, botanists do not make this distinction. Unlike zoologists, botanists use the term 'herbivory' more broadly to refer to antagonistic interactions between plants and the animals that eat them. But strict adherence to hard classifications of different types of species interactions is not wise. Textbook classifications of (e.g., predation, parasitism, competition, mutualism) are best viewed as signposts within a continuous field of species interactions. Upon closer inspection, species interactions often defy uniformly precise classification (Box 1.4).

BOX 1.4 **Species interactions**

Species can interact in a myriad of different ways. This diversity has traditionally been described by five distinct categories of species interactions, all of which are based on the effects each species has on the another's fitness. Predation occurs when one species benefits and the other is harmed, fatally. Parasitism occurs when one species benefits and the other is harmed without causing mortality. When both species are harmed in the interaction it is known as competition and when both species benefit it is mutualism. When species interact with no net effect on the other's fitness it is known as commensalism.

While helpful superficially, distinct categorisations of species interactions fail to capture the rich complexity of species interactions that actually take place in nature. In reality, the effect one species has on another usually defies rigid classification and is instead functionally continuous (Fig. B1.4).

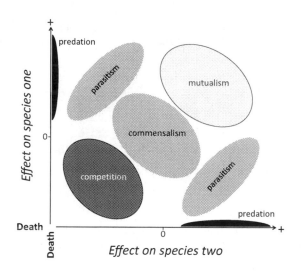

FIGURE B1.4 A diagram illustrating the continuous nature of different types of species interactions. The x- and y-axes depict the fitness effect of one player on the other, running from maximally negative (i.e., death) at the origin to increasingly positive effects at higher values.

BOX 1.4 **(cont.)**

Imagine a seed dispersal mutualism wherein a frugivorous animal (e.g., a giant tortoise) consumes a fleshy fruit (e.g., a wild tomato). The journey through the digestive tract of an animal is fraught with difficulty. After all, it's designed to breakdown ingested food. Frugivores routinely kill a subset of the seeds they ingest and some types of frugivores destroy greater numbers of seeds than other frugivores (e.g., Díaz Vélez et al. 2017). At what point does passive seed predation by frugivores tip the balance from 'mutualism' to 'parasitism' (Duthie et al. 2006; Burns 2008; Morgan-Richards et al. 2008). From the mother plant's perspective, perhaps the loss of even a single seed to the digestive tract of a frugivore should be viewed as a negative interaction. However, plants can produce thousands of seeds throughout their lifetime, so the loss of some during the dispersal of others might not have a net negative effect of her reproductive fitness. On the other hand, from a seed's perspective, such an interaction would be predation. The terms competition, predation, parasitism, mutualism, and commensalism might therefore be better conceptualised as regions in a complex landscape of continuous variation in species interactions.

Anagenesis versus Cladogenesis

Perhaps the most renowned giant tortoises on the planet are those that inhabit the Galapagos Islands. They occur on many of the archipelago's larger islands, with some islands housing multiple species. As vice governor Lawson was so keen to point out, shell shape can differ markedly among species. Yet, all of the giant tortoises inhabiting the Galapagos evolved from a single common ancestor, despite their differences in distribution and appearance (Caccone et al. 2002). They are therefore an example of *cladogenesis*, or the evolution of multiple taxa from a single common ancestor.

Giant tortoises are also an example of *adaptive radiation*, a special type of cladogenesis (see Givnish 2015). Adaptive radiation is specifically defined as the evolution of phenotypic and taxonomic diversity within a rapidly evolving lineage (Schluter 2000). Previous work on island evolution has focused disproportionately on lineages undergoing cladogenesis, and in particular on adaptive radiations. This book will take a different approach. It will focus on *anagenesis* or the gradual evolutionary transformation of a single branch of an evolutionary lineage into a new species. This form of speciation is perhaps best illustrated by our previous emblematic animal, the takahe. Instead of multiple species evolving from a single common ancestor, both *Porphyro* species in New Zealand evolved independently along single branches on their shared evolutionary tree.

Research Philosophy & Scientific Methodology

> ... the study of oddities became the approach to island biology for a long time. Carlquist's Island Biology (Carlquist 1974), in which he identified a suite of phenomena peculiar to island organisms is a culmination of this approach ... One may call it "the dodo approach". This approach will reveal interesting aspects of the endemics on any island. On the other hand, the majority of the endemic organisms may look quite similar to organisms in similar habitats in continental areas, and it is still undocumented that there is a larger proportion of oddities among island endemics ...
>
> Adsersen (1995), p. 8

Sherwin Carlquist is, quite clearly in my opinion, one of the great natural historians of all time. Most of the topics covered in this book were pioneered by his painstakingly astute observations (see also Traveset et al. 2016a). However, the research philosophy and scientific methodology adopted here will be markedly different.

Carlquist's (1974) methods followed tradition in natural history. Hours of curiosity-driven field observations resulted in many astounding examples of particular island phenomena. My favourites include fruit morphology in the genus *Bidens* (Asteraceae) in the South Pacific, which illustrate the loss of dispersal and the evolution of seed gigantism; stem and flower morphology in *Brachysema daviesioides* (Fabaceae) in Southwest Australia (an insular region of Australia according to Carlquist [1974; 2009]), which illustrate striking coadaptation to unusual herbivores and pollinators; and the morphological diversity exhibited by the genus *Cyanea* (Campanulaceae) in Hawai'i, which is arguably the best example of an adaptive radiation in plants and illustrates an almost unbelievable diversity of morphology and size.

Scientists interested in island biology responded to Carlquist's work in markedly different ways (see Midway & Hodge 2012). The first was to believe that his examples were indicative of more widespread patterns – repeated patterns in evolution that characterise most, if not all, insular biotas. In other words, the changes in fruit morphology exhibited by *Bidens* spp. (i.e., shortened awns and bigger seeds) could be extrapolated to characterise other, as yet unstudied, island plants. When viewed from this rather overzealous perspective, the search for an island syndrome in plants was over before it started. The island syndrome was real. Perhaps without knowing it, Carlquist (1974) thereby sparked a scientific paradigm shift (*sensu* Kuhn 1962) in how these readers thought about island biology.

Another response was to give natural history a bad name. To those adhering to the *hypothetico-deductive* approach to science (Popper 2005), Carlquist's examples, however elegant they might be, were nothing more than qualitative descriptions of particular species. As such, they may or may not characterise other island taxa. Until seed and awn sizes were measured in other island taxa and compared to analogous data from their mainland relatives, Carlquist's observations of fruit morphology in *Bidens* were merely anecdotes.

Philosophically, the hypothetico-deductive method argues against extrapolating from one or a small number of observations to encompass wider circumstances. Instead, it argues that scientific understanding proceeds by repeatedly testing and rejecting hypotheses, which can then be refined to reflect natural phenomena more accurately (Popper 1959). According to this research philosophy, belief in any paradigm is unwise.

This book adopts the hypothetico-deductive research philosophy wholeheartedly (cf Midway & Hodge 2012). The science of ecology has grown enormously in the last 50 years, and the traditional, descriptive approach is now as extinct as the dodo. But this book will not abandon Carlquist's work, quite the opposite. Most chapters in this book will begin with Carlquist's natural history genius, along with the observations of his predecessors, most notably Charles Darwin. But instead of extrapolating outwards from Carlquist's observations uncritically, his observations will be reformulated as hypotheses, which will then be tested empirically, both with new analyses and by reviewing the literature.

This will not be the final word on the search for an island syndrome in plants. True to the hypothetico-deductive research philosophy, each chapter will attempt to either falsify the existence of a repeated pattern in evolution, or fail to falsify it. The former will illicit the development of new hypotheses. The latter will leave the door open for continued testing. Hopefully, this approach will help transform the study of island biology methodologically from the 'dodo approach' into a rigorous, quantitative branch of scientific enquiry.

Giant tortoises provide an excellent illustration of the dangers associated with extrapolating beyond a handful of insightful observations. True to their name, giant tortoises are huge when compared with their closest living relatives. Similar to the takahe and pukeko, molecular analyses have been used to pinpoint their closest living relatives, all of which live on continents and are comparatively diminutive (Caccone et al. 1999; Austin et al. 2003; Le et al. 2006). Therefore, giant tortoises could be an example of parallel evolution towards island gigantism (Jaffe et al. 2011).

Recent paleontological discoveries tell a very different story. Fossil remains of giant tortoises are actually quite common on continents (Rhodin et al. 2015). These findings suggest that at least some giant tortoise lineages on islands were derived from giant-sized colonists from the mainland (Pérez-García et al. 2017). Given the widespread extinction of giant tortoises on continents, phylogenetic analyses are left to identify more distantly related, smaller-bodied extant species as their closest mainland relatives. Therefore, gigantism does not appear to be a derived characteristic of island tortoises. Large body size likely evolved on the mainland prior to their subsequent colonisation of islands.

FLORES ISLAND 'HOBBIT'

Flores Island is one of many, similarly shaped islands in the Indonesian Archipelago, sandwiched between Lombok and Sumbawa in the west, Sumba in the south, and Timor in the east. First impressions from a quick look at a map suggest that it would not be a very good place to find an emblematic island animal. It just does not seem isolated enough. But insularity is a continuously distributed characteristic of geography and first appearances can be deceiving.

During the Last Glacial Maximum (LGM: 18,000–32,000 years ago), huge amounts of water were locked up at the Earth's poles. As a result, global sea levels during the LGM were much lower than they are today, and many islands that are now separated by narrow channels of water were connected to nearby landmasses. Given the narrow straits of water that currently surround Flores Island, it's tempting to conclude that it was recently connected to nearby islands, making it far less 'insular' from a historical perspective. However, a different story emerges from a closer inspection of the bathymetry surrounding the island. The narrow channels of water that encircle Flores Island are actually quite deep, so when sea levels dropped during the LGM, Flores Island retained its insularity (Silver et al. 1983).

Among the animals that presently call the island home, only one has characteristics that are consistent with the island syndrome. The Flores Island giant rat (*Papagomys armandvillei*: Muridae) grows to half a metre long – about twice the size of the more familiar Norwegian roof rat (*Rattus norvegicus*). Similar to takahe and giant tortoises, overhunting by humans has reduced its numbers severely, so it is now quite rare to see in the wild.

Recall that mammals tend to be poor island colonists, probably because their demanding metabolic requirements limit survival during transit. Therefore, it's somewhat surprising that this example of island evolution has fur, rather than feathers or scales. But because Flores Island is separated from its neighbours by slender fingers of water, some of which are less than 30 kilometres wide, this particular test of mammalian survival during dispersal is not that lengthy. So, at some point in the evolutionary past, it would appear that rats on nearby landmasses were swept out to sea on flotsam, which then washed ashore on Flores after a relatively short journey at sea. However, a single species of giant rat is just the tip of an evolutionary iceberg that rests just below the surface of time.

Delving deeper into the evolutionary history of Flores Island reveals that it was a very, very different place in the recent geologic past. Without the unlucky animals who died in a series of magnificent limestone caves long ago, we would know nothing about the fauna that previously called the island home. One archaeological site in particular, known as Liang Bua (Fig. 1.4), contains the bones of a wonderful lost world of biodiversity that seems to have been shaped

FIGURE 1.4 Excavations in Liang Bua cave, Flores Island, Indonesia. (photo taken by Brent Alloway)

by insularity (Van den Bergh et al. 2009). 'Spectacular' would under-rate its contents. The sediments beneath the floor of Liang Bua conceal a treasure trove of subfossilised bones that have brought anthropology on a collision course with island biology (Fig. 1.4).

It turns-out that the extant giant rat was not alone. Two other species of giant rat previously inhabited the island (Locatelli et al. 2012). One was closely related (*Papagomys theodorverhoeeveni*), while the other was quite different (*Spelaeomys florensis*), and likely resulted from a separate colonisation event.

The bones of many different types of birds that can still be seen on the island today, including ducks, pigeons, rails, owls, and honey-eaters, are buried just beneath the surface (Van den Bergh et al. 2009). The presence of these familiar birds indicate that the island has retained much of its avifauna. However, there is one glaring exception. By far the most remarkable bird bones unearthed to date from Liang Bua come from a species of giant, flightless stork, which was diametrically different to how storks are portrayed in popular culture, as gentle, newborn delivery boys. This bird was the stuff of night-mares. Standing over two meters tall, the carnivorous Flores Island giant stork (*Leptoptilos robustus*: Ciconiidae) could have easily made a meal of giant rats or even bigger animals.

L. robustus was clearly flightless – relative to its body size, its wings were far too small to carry it into the air. Similar to the takahe, it's a fine example of insular flightlessness. When considered along-side the takahe, dodo, and kakapo, *L. robustus* is also an example of allopatric convergence towards insular gigantism. However, not all island animals evolve into giants. It turns out that the opposite trend, insular dwarfism, may be equally common on islands.

Paradoxically, while some types of animals evolve into giants on islands, other types of animals evolve into dwarfs. For example, the world's smallest bird (the Cuban bee hummingbird, *Mellisuga hele-nae*: Trochilidae), lizard (the Malagasy leaf chameleon, *Brookesia micra*: Chamaeleonidae), and snake (the Barbados threadsnake, *Lepto-typhlops carlae*: Leptotyphlopidae) live in insular environments.

Insular dwarfism is also common in mammals, particularly ungulates (huffed animals) and carnivores (canids, felids, and mustalids).

The lost fauna of Flores Island also contained dwarfs. The bones of several dwarfed elephants in the genus *Stegodon*: Proboscidea (Van den Bergh et al. 2009) are commonly unearthed at Liang Bua and other older sites on Flores (e.g., Soa Basin, see Brumm et al. 2016). Similar excavations on islands in the Mediterranean Sea, Bering Sea, and off the coast of California have produced more dwarfed proboscideans that evolved convergently with the Flores Island *Stegadon* (van der Geer 2010; Rick et al. 2012).

The Island Rule

This paradox of body-size evolution, with some animals evolving into dwarfs, while other animals evolving into giants, has a name, the *island rule*. The island rule is a putative pattern in body-size evolution, wherein small animals evolve to become larger on islands, while large animals evolve to become smaller (Box 1.5). Even though it makes no specific predictions *per se* about the processes underpinning body-size evolution on islands, the island rule has sparked a long history of significant research interest, as well as controversy and debate (Meiri et al. 2011; Lomolino et al. 2012; Itescu et al. 2014; Faurby & Svenning 2016).

At the heart of the island rule debate is whether the direction of body-size evolution on islands is related to body size itself. Research on a diverse range of animals, including reptiles, mammals, and birds, has shown that evolution produces both giants and dwarfs on islands. The question is whether these changes can be predicted by mainland body sizes. Despite the enormous body of literature that has attempted to answer this question, whether plants might follow the island rule has yet to be explored.

In 2002, ongoing excavations in Liang Bua unearthed a truly remarkable and unprecedented dwarfed mammal – a species that would bring the topic of island biology to our own evolutionary doorstep. Lying alongside the bones of other extinct animals, both

BOX 1.5   **The island rule**

One of the most distinctive aspects of island faunas is their body size. Paradoxically, both giant rats and pigmy mammoths evolved together on Flores Island in Indonesia, as well as several other islands across the globe (Rick et al. 2012). Similarly, giant lemurs and dwarf hippopotamuses evolved together in Madagascar, and islands in the Mediterranean housed both dwarfed deer and giant hedgehogs (van der Geer et al. 2010).

In 1964, a mammologist named Berry Foster who was working in the Haida Gwaii Islands off the Northwest Coast of North America, noticed a trend in the size of island animals, which has subsequently become known as the island rule (Foster 1964). The island rule refers to a graded trend in body-size evolution wherein small-bodied species evolve to become larger on islands, and large-bodied species evolve to become smaller. More specifically, shrews, mice, and rabbits are predicted to evolve into giants, while foxes, deer, and elephants evolve into dwarfs. Couched in different terms, the island rule predicts that insularity tends to favour intermediate body size in animals (Fig. B1.5).

The island rule has since been the subject of intense empirical scrutiny and debate. Building upon an initial focus on mammals (see Meiri et al. 2008; Lomolino et al. 2012; Faurby & Svenning 2016), subsequent workers tested whether it might also apply to lizards (Meiri 2007; Meiri et al. 2011), turtles (Itescu et al. 2014), birds (Meiri et al. 2011), insects (Palmer 2002), and even dinosaurs (Benton et al. 2010). Results are not straightforward (Faurby & Svenning 2016; Lokatis & Jeschke 2018).

One mechanism that might give rise to the island rule is that evolution favours a reduction in interspecific variability in body size on islands (see Lomolino et al. 2012). Given that islands house more species than continents, interspecific competition for resources might be higher on the mainland. If so, increased competition on the mainland may promote displacement in mammal body sizes. Conversely, reduced levels of diversity on islands may result in the evolutionary convergence towards intermediate body size, thus promoting the island rule. See

BOX 1.5   **(cont.)**

Whittaker and Fernández-Palacios (2007) for a description of other mechanisms that may promote the island rule.

(a)

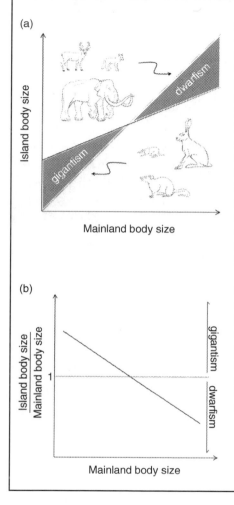

Island body size

Mainland body size

FIGURE B1.5A A graphical depiction of the island rule. Body size of island species (y-axis) is plotted against body size of mainland species (x-axis). The dashed line represents no evolutionary change in insular body size (isometry) and the solid black line represents a hypothetical maximum change in body size. The shaded region close to the origin represents gigantism (exemplified by insectivores, lagomorphs, and rodents), while the shaded region away from the origin represents dwarfism (exemplified by carnivores, proboscideans, and ungulates).

(b)

Island body size / Mainland body size

gigantism

dwarfism

1

Mainland body size

FIGURE B1.5B An alternative method of testing for the island rule. Insular size changes (the ratio of island size to mainland size) is plotted on the y-axis. Mainland sizes are plotted on the x-axis. A negative relationship between insular size changes and mainland sizes is predicted by the island rule.

big and small, were the bones of a creature from within our own evolutionary lineage. A dwarf hominin that was so unusual, yet at the same time so familiar, that it has permanently altered our view of human evolution (Fig. 1.5).

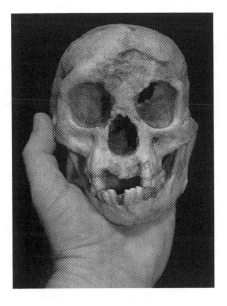

FIGURE 1.5 A skull of *Homo floresiensis* from Flores Island, Indonesia. (photo taken by Brent Alloway)

H. floresiensis: Hominidae was the scientific name given to remains of a number of individuals entombed within Liang Bua cave approximately 70,000 years ago (Brown et al. 2004; Morwood et al. 2004, 2005; Sutikna et al. 2016). The most complete specimen (LB-1) comprised a complete skull with dentition, as well as shoulder, arm, hip, leg, and foot remains – the morphology of which clearly indicated that it was a hominoid. In fact, the shape of its bones seemed so similar to ours that it appeared to belong in our own genus, *Homo*. However, it also had attributes that were different. It was strikingly small and had disproportionately large feet. Standing approximately one metre above the ground, it was similar in size to the smallest known species in the more 'primitive' genus *Australopithecus*, which evolved much earlier in Africa. These and other differences between our bones and the Flores Island 'hobbit', as it was dubbed by the popular media, seemed to delineate it from modern humans. So its discoverers labelled it as a new species.

The researchers who first discovered *H. floresiensis* hypothesised that its diminutive stature was consistent with insular dwarfism (Brown et al. 2004, see also Montgomery 2013). After all, we too are animals, and, like other animals inhabiting isolated islands, we are potentially subject to the same selection pressures. Going one step further, some argued that the unusual size of *H. floresiensis* was consistent with the island rule (Bromham & Cardillo 2007). Hominins as a group are relatively large mammals, so, according to the island rule, they should evolve small body size on islands, in much the same way as the Flores Island Stegodon species. However, this explanation for the unusual morphology of the Flores Island hobbit would prove contentious.

Soon after it received its new name, other anthropologists began to question whether *H. floresiensis* was really what it appeared to be. Was it really a new species of 'insular' hominin that rapidly evolved its diminutive stature after arrival in the insular environment of Flores Island (Van den Bergh et al. 2009)? Or was it actually something else? From an alternative perspective, many of the diagnostic features of the Flores Island hobbit (e.g., dwarfed stature and small cranium size) are also consistent with variety of developmental abnormalities that afflict humans. *Microcephaly* refers to a range of birth defects that result in dwarfed body size and shape (Argue et al. 2006; Martin et al. 2006; Syrmos 2011). If the microcephaly hypothesis were correct, then *H. floresiensis* would be a misnomer, as the bones would actually belong to members of our own species (*Homo sapiens*) and their dwarfed stature resulted from pathological problems, rather than insular selection pressures (Niven 2007).

Efforts were made to extract DNA from the hobbit's bones, in the hope of testing for genetic abnormalities associated with microcephaly. Its DNA could also be compared to similar samples of ancient DNA from other extinct species of *Homo* (e.g., *Homo neanderthalensis*; Noonan et al. 2006) to test the phylogenetic hypothesis depicted in Fig. 1.6. However, all attempts to isolate DNA from its subfossilised bones proved fruitless (see Jones 2011). DNA is a fragile molecule and the warm climate of Liang Bua, coupled with the subfossil bone's long tenure at the bottom of the cave, seems to have abraded all trace of its genetic signature.

Flores Island today is home to a range of people with different ethnic backgrounds. Interestingly, people in the village of Rampasasa, not far away from Liang Bua, are remarkably short in stature. Adults

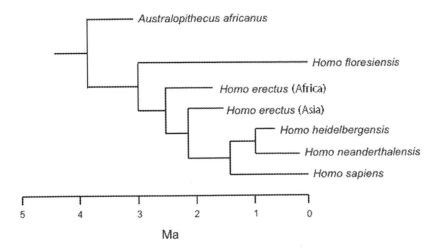

FIGURE 1.6 A phylogenetic hypothesis for the evolutionary history of the genus *Homo* based on dentition (Evans et al. 2016).

reach an average height of 145 cm, which is substantially smaller than most other ethnic groups. The average height of Americans from a diverse range of ethnic backgrounds is approximately 168 cm. Tucci et al. (2018) conducted a series of molecular analyses on this modern population of 'pigmies' and found that their diminutive stature is genetically determined and cannot be explained by developmental abnormalities or poor diet. Therefore, hominins may have walked the evolutionary pathway towards dwarfism multiple times on Flores Island in a manner similar to the taxon cycle.

Allometry & Ontogeny

Upon closer inspection, the island rule and microcephaly hypotheses actually make unique predictions regarding the Flores Island hobbit. Under the island rule hypothesis, differences between the size of the hobbit's bones and the bones of one of its closest mainland ancestors (*Homo erectus*, Fig. 1.6), should be consistent with evolutionary size changes in other primates that are endemic to isolated islands. On the other hand, the microcephaly hypothesis predicts that allometric relationships between brain and body size should match those of

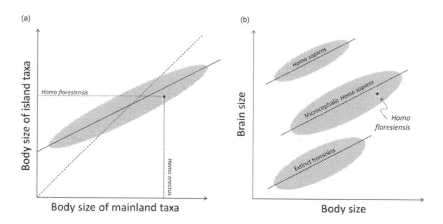

FIGURE 1.7 (a) Hypothetical relationship between the body size of primates inhabiting isolated islands and their closest mainland relatives. The grey oval encapsulates all species pairs and the solid line represents the best-fit relationship, as predicted by the island rule. This hypothetical relationship between *Homo floresiensis* and *Homo erectus*, a closely related mainland taxon, is placed within the region of observed values for primates, providing support for the island rule. (b, top) relationship between brain size and body size in developmentally normal *Homo sapiens*, (b, middle) microcephalic *H. sapiens*, and (b, bottom) extinct hominins. If *H. floresiensis* were to fall within the region occupied by microcephalic *H. sapiens*, it would be consistent with the microcephally hypothesis.

microcephalic humans, rather than developmentally normal humans or other extinct hominins (Fig. 1.7).

Exploring relationships between body size and brain size in the Flores Island hobbit and its close relatives elsewhere has proved quite popular. Many allometric analyses have been conducted and to date most are inconsistent with the microcephally hypothesis (Falk et al. 2005; Argue et al. 2006; Gordon et al. 2008; Lyras et al. 2009; Montgomery 2013; Brumm et al. 2016; Evans et al. 2016; Van den Bergh et al. 2016; Argue et al. 2017; but see Martin et al. 2006). Therefore, at present, the Flores Island hobbit is best referred to as *H. floresiensis*, a dwarfed island relative of *H. erectus*.

Analyses of allometry can be an important tool in island biology. Animals often change shape as they develop, and these

BOX 1.6 **Allometry & ontogeny**

Covariance between the size and shape of morphological traits is generally known as allometry. Allometric relationships between traits can arise from a variety of different processes. Natural selection can promote trait covariation directly if different traits work together to fulfil a particular purpose. For example, leaf and stem sizes often covary allometrically, because bigger leaves require greater mechanical support (Fig. B1.6a; Ackerly & Donoghue 1998; Olson et al. 2009). On the other hand, webs of developmental, genetic, and physiological constraints can also promote trait correlations indirectly (see Armbruster et al. 2014). For example, the size of leaves and fruits can vary allometrically (Fig. B1.6b), even though the two traits do not appear to be linked functionally. So, if selection acts on leaf size, it may affect the evolution of fruit size indirectly (Herrera 2002a).

The relative size and shape of traits can also vary substantially during development (i.e., ontogeny). Ontogenetic changes in plant morphology, such as changes in leaf size or shape (Fig. B1.6.b), are commonplace and known as *heteroblasty*. Several processes may select for heteroblasty (see Zotz et al. 2011). As plants grow upwards, away from the ground surface, environmental conditions change. So, heteroblasty may be selected for by changes in environmental conditions, such as light, if 'juvenile' traits are better adapted to shaded conditions on the forest floor and 'adult' traits are better adapted to sunnier conditions in forest canopies. Heteroblastic changes in morphology often occur continuously throughout development, leading to gradual changes in morphology between juveniles and adults. On the other hand, heteroblasty can also be abrupt, with pronounced morphological changes occurring rapidly during a brief window in ontogeny. For example, plants growing in aquatic environments often shift from producing filamentous leaves when they are underwater, to flattened, floating leaves once they reach the surface of the water column (Zotz et al. 2011). This abrupt form of heteroblasty is analogous to metamorphic development in some types

BOX I.6   (cont.)

of animals, such as lepidopterans that transform from caterpillars to butterflies suddenly during ontogeny.

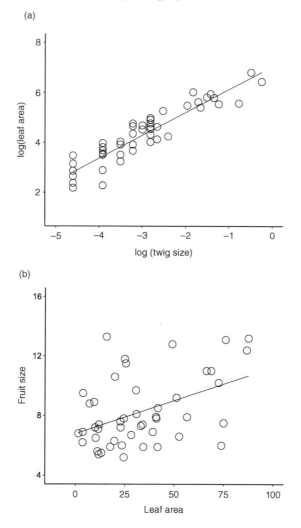

FIGURE B1.6A (a) Allometric relationships between leaf area and twig size in 48 deciduous trees from eastern North America (data from White 1983). (b) Allometric relationships between fruit size and leaf area in 42 species of fleshy-fruited plants from the Iberian Peninsula (data from Herrera 2002b).

BOX 1.6 (cont.)

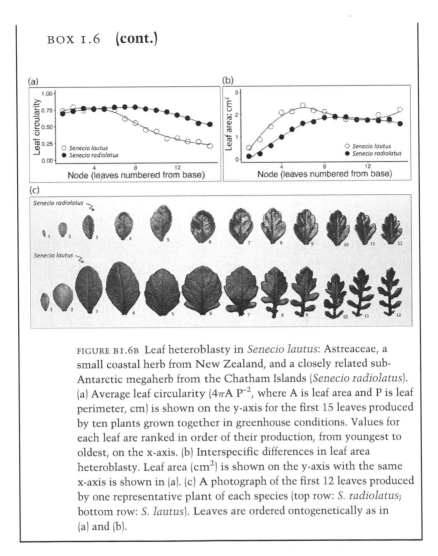

FIGURE B1.6B Leaf heteroblasty in *Senecio lautus*: Astreaceae, a small coastal herb from New Zealand, and a closely related sub-Antarctic megaherb from the Chatham Islands (*Senecio radiolatus*). (a) Average leaf circularity ($4\pi A\ P^{-2}$, where A is leaf area and P is leaf perimeter, cm) is shown on the y-axis for the first 15 leaves produced by ten plants grown together in greenhouse conditions. Values for each leaf are ranked in order of their production, from youngest to oldest, on the x-axis. (b) Interspecific differences in leaf area heteroblasty. Leaf area (cm^2) is shown on the y-axis with the same x-axis is shown in (a). (c) A photograph of the first 12 leaves produced by one representative plant of each species (top row: *S. radiolatus*; bottom row: *S. lautus*). Leaves are ordered ontogenetically as in (a) and (b).

ontogenetic changes can help them to adapt to novel island environments. Plants are no different and an explicit consideration of allometric changes in size through ontogeny is often needed when characterising how island plants differ from their mainland relatives (Box 1.6).

Regardless of the methods used to investigate the size of island animals, size changes are a common feature of island faunas. Curiously, an analogous literature on plants does not exist. Chapter 5 seeks to fill this gap by exploring size changes in island plants, and, in particular, testing whether plants obey the island rule.

CONCLUSIONS

There are many pathways to scientific discovery. No one way is 'correct'. However, a critically important ingredient to scientific progress is questioning popular dogma. In 1920, popular opinion argued that the takahe was extinct. However, Geoffrey Orbell was sceptical of this situation, and ultimately successful at locating some of the last surviving adults. Superficially, giant tortoises seem like a fantastic example of insular gigantism. However, by digging a little bit deeper into the problem, palaeontologists have discovered that they were previously common on continents, so their big bodies might not result from insular selection pressures. Ongoing study of the dwarfed stature of hominins on Flores Island will hopefully paint a clearer picture of our own evolutionary past, as well as the developmental circumstances and timeframe required for insular body-size evolution to occur in general.

The remainder of this book is built upon the foundations of Sherwin Carlquist's natural history genius. Nevertheless, it openly questions the generality of his observations by adopting the hypothetico-deductive scientific method. Five aspects of the life history of island plants are explored in separate chapters, most of which have a clear precedent in island animals. The final chapter concludes with discussions of five emblematic island plants that exemplify the evidence for an island syndrome in plants. These five species also illustrate a range of life history attributes that may be part of an island syndrome, but require future study, in addition to several attributes that have traditionally been included in an island syndrome, but failed to pass the test.

2 Differences in Defence

Islands often house very different assemblages of predators than continents. Given their poor colonisation potential, mammalian predators are often replaced by other types of predators with greater dispersibility, such as birds. As a result, prey species on islands regularly lose defences against predators that are absent from islands, and readjust their anti-predatory adaptations to suit those that are present.

Are island plants similar to island animals? When freed from attack by mainland herbivores, do island plants lose defensive adaptations commonly deployed by their continental counterparts? On islands housing unusual 'toothless' herbivores, do island plants consistently evolve traits to cope with them?

Two types of data are needed to answer these questions. First, herbivore occurrence patterns need to be established to determine which islands were colonised by particular types of herbivores and which ones were herbivore-free. Second, the effects of particular plant traits on the behaviour of different types of herbivores need to be determined. Both types of data can then be integrated to test the hypothesis that island plants consistently adjust their defensive adaptations to suit co-occurring herbivore faunas. Straightforward as it seems, how plant defensive adaptations evolve on isolated islands is infrequently documented, poorly understood, and, in some cases, highly contentious.

ISLAND HERBIVORES

Insects are arguably the most abundant and widespread herbivores on the planet (Strauss & Zangerl 2002). They have co-evolved with plants for 400 million years and have had a significant influence on the evolution of plant form and function. However, insect herbivores will not be the focus of this chapter, both because their biogeographic distributions are comparatively difficult to establish relative to vertebrates, and because trait-based relationships between plants and particular types of insects are more complicated. For the same reasons, a wide assortment of other types of herbivores, including viruses, bacteria, fish and fungi, await further study. Instead, this chapter will focus on terrestrial vertebrate herbivores.

Mammals are the largest vertebrate herbivores on the planet and have similarly profound impacts on plant ecology and evolution to insects (Danell & Bergström 2002). Not surprisingly, many plants possess adaptations to deter mammalian herbivores (see Hanley et al. 2007). When mammal-defended plants initially colonise islands from the mainland, founding populations carry these defences with them. But because mammals are poorly equipped for overwater dispersal, they rarely occur on isolated islands and plant defences against mammals lose their usefulness. On the other hand, plants colonising many large, oceanic islands were greeted by avian and reptilian herbivores. Although relatively unimportant on most continents, these 'toothless' herbivores dominated many island ecosystems prehistorically.

Unfortunately, vertebrate herbivore faunas have undergone radical transformations in the Anthropocene, not just on continents, but especially on islands (Steadman 2006). Native vertebrate herbivores have been eliminated from many islands and replaced intentionally with continental herbivores. So the herbivore faunas that occur on islands today are often a poor reflection of the circumstances under which island floras evolved.

Given their recent disappearance, identifying repeated patterns in plant defence against native vertebrate herbivores requires an understanding of prehuman herbivore distributions, which can be acquired in several ways. Some isolated islands were discovered relatively recently by peoples with written language. For many of these islands, such as the Galapagos, Mauritius, and Socotra, there are written descriptions of island faunas at the time of human contact. Other islands were discovered by peoples without written language and lack written descriptions of the animals they once housed. Under these circumstances, paleontological excavations of subfossil bones, similar to those in Liang Bua on Flores Island, are needed to establish the types of herbivores that were once present.

Tortoises

Eighty Ma, reptiles were the most important vertebrate herbivores on the planet. But, since then, reptiles have declined in importance as herbivores, especially after the extinction of dinosaurs 65 Ma. Giant tortoises, the largest-bodied herbivorous reptiles alive today, can no longer be found on continents. However, a very different situation emerges on some oceanic islands, where giant tortoises were exceedingly common prehistorically. Early European explorers were astounded by the enormous densities of tortoises on the islands where they occurred (Cheke & Hume 2008). Given their numerical dominance, native plants may have evolved specific defences against tortoise herbivory.

Several aspects of the morphology and physiology of giant tortoises place clear constraints on how they interact with plants. First, tortoises do not have teeth, so they cannot chew their food. Instead, they must swallow their food more or less whole. As a result, they are likely to have difficulty consuming stiff, oversized, or spinescent leaves. They also have keratinised bills instead of sensitive lips and gums, so their first point of contact with food plants is more impervious to damage than mammalian mouthparts.

Second, reptiles can see across a broader spectrum of wavelengths than mammals. Most herbivorous mammals (e.g., ungulates) have just two light receptors in their eyes, which limits the wavelengths of light they can use to discriminate their food (Jacobs et al. 1998). On the other hand, reptiles possess four light receptors and are therefore able to detect a wider range of wavelengths (Bowmaker 1998). As a result, tortoises are more likely to rely on visual cues to locate palatable plants; therefore, selection could favour plants with traits that make them more difficult to detect visually.

Third, giant tortoises have a limited capacity to access leaves produced above the forest floor, given their heavy bodies and thick shells. So, leaves and shoots located higher up in plant canopies are free from danger. Plant defences against tortoises are therefore likely to be deployed preferentially by small-statured plants and during early ontogenetic stages in taller-statured plants.

Birds

With only a handful of exceptions (e.g., the hoatzin, *Opisthocomus hoazin*: Opisthocomidae), strongly herbivorous birds have given up the ability to fly for several reasons. Plants are generally less nutritious than animal prey, so herbivores need to consume larger quantities of foliage to survive. The weight of this foliage adds to the mass of birds during digestion, which negatively impacts their capacity to fly. Plant leaves and stems also require prolonged processing times, which can only be achieved in long digestive tracts, which are necessarily characteristic of large-bodied species. Weighed down by their herbivorous lifestyle, leaf-eating birds tend to be large bodied and, in most cases, have become flightless.

Several species of herbivorous ratites currently inhabit continents in the southern hemisphere, including the ostrich (*Struthio camelus*: Struthionidae) in Africa, rhea (*Rhea* spp.: Rheidae) in South America, and emu (*Dromaius novaehollandiae*: Casuariidae) in Australia. Their comparatively low diversity and restricted distribution makes them relatively unimportant herbivores on a global scale. However, prior to the wave of extinctions at the outset of the Anthropocene, giant birds were important herbivores on many oceanic islands. Most of these species colonised islands as small-bodied omnivores and subsequently evolved into giant, flightless browsers (e.g., Mitchell et al. 2014; Yonezawa et al. 2017). Lineages that have travelled down this evolutionary pathway towards herbivory include Galliformes (Sylviornithidae) in New Caledonia, water fowl (Anatidae) in Hawai'i and ratites (Palaeognathae) in Madagascar and New Zealand.

Despite their differences in appearance, giant tortoises and herbivorous birds actually forage in similar ways using analogous mouth parts. Both taxa lack teeth, so they cannot chew and they have keratinised bills instead of soft lips, so they are less susceptible to damage at the initial point of contact with food plants. With four visual receptors, they both see well and are likely to be more visually orientated foragers than mammals. Because they can no longer fly, they are also restricted to foraging from the ground, so foliage located higher up in plant canopies is out of reach.

Islands without Vertebrate Herbivores

Although many large, oceanic islands housed toothless herbivores, smaller islands often lacked vertebrate herbivores altogether, perhaps because small islands were insufficient in area to sustain viable populations (MacArthur & Wilson 1967). Plants on herbivore-free islands are released from selection by vertebrate herbivores altogether, regardless of which ocean they occur in, or from where they accumulated their respective floras.

DETERRENTS TO HERBIVORY

Chemistry

Plants utilise a variety of chemical compounds to deter herbivores and protect leaves (Agrawal & Weber 2015). One of the most common are tannins, which bind to digestive enzymes and dietary proteins, thus inhibiting their functioning (see Farmer 2014). Cyanogenic glycosides are another type of toxic secondary metabolite, which, when consumed by herbivores, is converted by enzymatic activity into hydrogen cyanide, an extremely potent poison (Vetter 2000). Alkaloids are a heterogeneous group of chemicals including caffeine, nicotine and strychnine that can have a variety of deleterious effects on animals (see Fester 2010; Walters 2011). Conifers often produce resins in response to attack, which can contain terpenes that are toxic to many types of herbivores (Krokene et al. 2010; Thimmappa et al. 2014).

A diverse range of plants also secrete latex when wounded, which can interfere with the foraging behaviour of herbivores, as well as contain toxic secondary chemicals (see Huber et al. 2015).

Spinescence

Another deterrent to vertebrate herbivory is sharp, ridged projections on or around vulnerable plant tissues. Spinescence can arise developmentally from several different plant tissues, each with its own terminology. Spines are modified leaves or stipules, thorns are modified twigs or branches, and prickles are derived from epidermal tissue. Regardless of their anatomical origin, prickles, thorns, and spines are functionally convergent traits that influence the behaviour of vertebrate herbivores (Cornelissen et al. 2003). Instead of deterring herbivory completely, spinescence tends to limit damage by large herbivores by reducing harvesting rates (Wilson & Kerley 2003a, 2003b; Cash & Fulbright 2005; Hanley et al. 2007; Shipley 2007).

Spinescence is often argued to be more effective in deterring mammalian browsers than avian browsers, which seems logical given that mammals have soft lips and gums that can be damaged more easily by sharp plant parts (Greenwood & Atkinson 1977; Bond & Silander 2007; Lee et al. 2010). However, some particular types of spinescence could be effective at deterring toothless browsers. Because birds and tortoises lack teeth and cannot chew, they cannot grind up food prior to ingestion. Therefore, leaves with sharp, rigid spines projecting from their margins may be difficult for toothless herbivores to swallow whole, and thereby constitute an effective deterrent to herbivory. Dense networks of thorns produced along the extremities of plant canopies (i.e., distally) could also protect leaves located within their inner recesses (i.e., proximally).

Branching Architecture

The way plants are branched might also deter herbivores. Most woody plants produce new branches at relatively shallow angles. As a result, their growth can be concentrated in particular directions, most often upwards. However, *divaricate* plants produce branches at consistently wide angles, leading to less directional growth and a distinctive wiry appearance.

The evolutionary origins of divaricate branching are an enigma but it has evolved independently in many lineages. One explanation is that the growth form evolved to enhance physiological performance, or to help plants cope with past climatic conditions (see McGlone & Webb 1981; Howell et al. 2002). Alternatively, it could have also evolved as a deterrent to vertebrate browsers (Carlquist 1974; Greenwood & Atkinson 1977; Atkinson & Greenwood 1989; Bond & Silander 2007).

Divaricate branching may protect plants in several ways. First, the twigs of divaricate species are often strongly reinforced, increasing the effort required to remove leaves (i.e., high tensile strength, see Bond et al. 2004a). Second, branches that are attached at right angles stretch out over greater distances when pulled from their tips, further increasing the energy required by browsers to remove their leaves with a plucking motion. Third, right-angled branches, which give rise to a 'zig-zag' appearance, would make larger segments of stems difficult to swallow whole (Bond & Silander 2007). Lastly, branches at the extremities of divaricate plants often have lower leaf densities than branches towards their centre. Therefore, the outer branches themselves could impede access to a majority of leaves produced by divaricate plants (Atkinson & Greenwood 1989). In some species, these outer branches are very rigid and have sharpened tips, which might further prohibit access to leaves located in their inner recesses (i.e., 'porcupine plants', Burns 2016a, see also Charles-Dominique et al. 2017).

Leaf Heteroblasty

Defences are often deployed differentially throughout a plant's lifetime (Zotz et al. 2011). For example, many trees are better defended chemically at early stages of development, when they are still within reach of vertebrate herbivores (see Swihart & Bryant 2001). Similarly, spinescent plants tend to produce prickles, thorns, and spines preferentially at early ontogenetic stages, and many divaricate plants lose their wiry appearance when they grow above a certain height (see Dawson 1988; Zotz et al. 2011; Burns 2013a).

Leaf morphology can also be heteroblastic. Divaricately branched trees always produce smaller leaves at earlier stages of ontogeny. Although juvenile microphylly could have a physiological explanation, it could also have been selected for by herbivores (Bond et al. 2004a). By reducing the energetic return to browsers that harvest leaves individually using a plucking motion, small juvenile leaf sizes might show rates of herbivory.

Leaves might also be coloured in ways that help protect them from attack (Lev-Yadun 2016). Some plants produce non-photosynthetic pigments in their leaves that radically alter their appearance and could make them more difficult for herbivores to locate. An object's conspicuousness results from a combination of its own reflectance properties and the reflectance properties of the background in which it lives. So, the more similar an object appears to its surroundings, the more difficult it is to locate. *Crypsis* refers to phenotypic similarity between a palatable object and unpalatable, inanimate objects, such as rocks or leaf litter (Kellner et al. 2011). *Mimicry* is a conceptually similar phenomenon. However, in this instance, evolution favours phenotypic similarity between a palatable object and a less-palatable, living organism (Barlow & Wiens 1977; Gianoli & Carrasco-Urra 2014; Scalon & Wright 2015; but see Blick et al. 2012). *Aposomatism* refers to the opposite phenomenon, when structural defences such as spinescence are conspicuously coloured to advertise their presence (Lev-Yadun 2016). If leaf colours evolved to thwart flightless vertebrate herbivores, we might expect colour-based defence to be heteroblastic in trees that grow above animal reach at maturity.

In addition to leaf size and colour, variability in the morphology of leaves can be more pronounced at early ontogenetic stages, perhaps for defensive purposes. Animals often forage by developing a 'search image' for appropriate food resources (Tinbergen 1960; Punzalan et al. 2005). Pronounced variability in size, shape, colour, or symmetry of leaves could prohibit a herbivore's ability to form a search image for palatable prey and thereby form a viable defence (Brown & Lawton 1991; Lev-Yadun 2016). Although few studies to date have tested the search image disruption hypothesis in plants, Dell'Aglio et al. (2016) found that *Heliconius* butterflies use leaf shape as a cue for foraging and oviposition, driving negative frequency dependent selection on leaf shape in *Passiflora* vines. This could potentially also apply to ground-dwelling, vertebrate herbivores.

Plant Protection Mutualisms

An alternative strategy of plant defence is to develop mutualistic associations with animals that protect the plant from harm. For example, many plants (e.g., *Acacia* spp: Fabaceae) provide shelter in the form of hollow stems or thorns for stinging ants to inhabit. Ants that inhabit these structures then protect their homes from attackers. Other plants provide food for ants, in the form of glycogen-rich Müllerian bodies, or extra-floral nectaries that secrete sugary secretions. Plant protection mutualisms can be an extremely effective plant defence and may involve a variety of arthropods, including ants and mites (Beattie 1985; Rico-Gray & Oliveira 2007; O'Connell et al. 2010; Ward & Branstetter 2017).

SYNDROME PREDICTIONS

Immigrants to islands often come from larger landmasses where their ancestors were subject to vertebrate herbivores, mammalian or otherwise. If the newly colonised island is devoid of vertebrate herbivores, they may often possess defunct defensive adaptations. Assuming defensive adaptations are costly to produce, and if resources previously devoted to vertebrate defences can be diverted elsewhere (e.g., growth, reproduction), evolution will favour the loss of defence (Lahti et al. 2009). This prediction can be clearly illustrated by a Venn

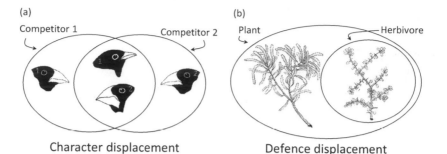

FIGURE 2.1 Venn diagrams illustrating character displacement and defence displacement. (a) Character displacement predicts that competing species diverge morphologically in sympatry to avoid competition for resources (see Schluter 2000; Chase & Leibold 2003). This is illustrated by two species of seed-eating finch (labelled 1 and 2), which have similar morphologies when they occur in allopatry, but in sympatry they evolve differences in bill morphology that promote differences in the use of food resources. Species 1 evolves a small bill to specialise on small seeds, while species 2 evolves a large bill to specialise on large seeds. (b) Defence displacement predicts that plant defences are deployed only in sympatry with herbivores. The large ellipse illustrates the total spatial distribution of a hypothetical plant species. The small ellipse illustrates the portion of its spatial distribution where it co-occurs with a hypothetical herbivore. The ellipse representing the herbivore sits entirely within the ellipse representing the plant, because the herbivore must co-occur with the plant in order to survive. Where the plant occurs sympatrically with the herbivore (e.g., close to the ground on continental landmasses), it evolves defensive adaptations for protection (e.g., thorns). However, in the absence of herbivore-mediated selection (either on isolated islands or when plants grow above the reach of mammalian herbivores on continents), plants are predicted to reinvest the energetic costs of defence elsewhere and become 'defenceless'. Illustration reprinted with permission from Burns (2013a)

diagram similar to those illustrating competitive displacement (i.e., *defence displacement*; Fig. 2.1).

Alternatively, defensive adaptations against toothless herbivores should evolve convergently on islands that housed giant tortoises or browsing birds. Assuming avian and reptilian herbivores forage analogously, three types of defensive adaptations might be more prevalent on 'toothless-browser islands'. First, spines along the edges of leaves should be more common than large thorns produced on woody stems (i.e., *insular spinescence*). Second, traits associated with divaricate branching, including high tensile strength, right-angled branches that have large displacement distances when pulled, and branching patterns that create a cage-like outer structure, should be more prevalent. Third, traits associated with leaf heteroblasty (e.g., microphylly, crypsis, mimicry, aposematism, and leaf-shape heterogeneity) should be produced preferentially at earlier ontogenetic stages.

Mutualisms only function when both players in the interaction co-occur in space and time. If a key partner is missing, the mutualistic relationship unravels because mutualistic services cannot be delivered. For example, in the absence of frugivores, the seeds in fleshy fruits will not be dispersed. In the case of plant protection mutualisms, the production of extra-floral nectaries, Müllerian bodies, or hollow stems will only be effective in deterring herbivory in the presence of stinging ants. Many isolated islands either have depauperate ant faunas or lack native ant mutualists. Under these circumstances, selection should favour the loss of mutualistic adaptations towards plant protectors and the dissolution of the plant protection mutualism.

HYPOTHESIS TESTING

Islands with Toothless Herbivores

Madagascar

Madagascar was attached to the ancient super-continent Gondwana 160 Ma. However, it began its geological journey towards insularity soon thereafter by rifting away into the Indian Ocean while still attached to India and the Seychelles (see Goodman & Jungers 2014). It became an island approximately 80 million years later and now lies 400 km east of Mozambique in the Indian Ocean.

The prehistoric herbivore fauna of Madagascar was truly exceptional. It housed two types of toothless browsers, both giant tortoises (*Aldabrachelys spp.: Testudinidae*) and 'elephant birds' (*Aepyornis* and *Mullerornis spp.: Aepyornithidae*). Unlike other islands that once housed toothless browsers, Madagascar was also home to multiple types of large mammalian browsers. Subfossils from several species of hippopotamuses have been discovered (*Hippopotamus lemerlei*, *Choeropsis madagascariensis: Hipopotamidae*), in addition to approximately 17 species of giant lemurs (e.g., *Megaladapis spp.: Megaladapidae*, *Archaeoindris spp.: Palaeopropithecidae*), many of which had strongly herbivorous diets (Goodman & Jungers 2014).

Malagasy plants display a variety of traits that may have protected them against toothless browsers. Bond and Silander (2007) recorded more than 50 species from 36 genera and 25 families that exhibit divaricate-like branching patterns. While similar in many ways to divaricately branched plants from New Zealand, Malagasy divaricates tend to produce greater quantities of leaves at their extremities, rather than concentrating leaf production in their inner recesses, as they tend to do in New Zealand. Bond and Silander (2007) therefore referred to them as 'wire plants' rather than divaricates. Comparisons with related species in Africa showed that they have thinner twigs, smaller leaves, and wider branch angles. Their leaves also have higher tensile strength and greater lateral displacement when pulled.

Many Malagasy plants are spinescent, although the incidence of spinescence in Madagascar is lower than in comparable habitats in Africa (Grubb 2003). Spinescence in many plant taxa (e.g., *Diospyros aculeata*: Ebenaceae; *Euonymopsis humbertii*: Celastraceae; *Rinorea spinose*: Violaceae) occurs along leaf margins, which is consistent with the insular spinescence hypothesis. However, many other spinescent plants produce large, widely spaced thorns on branches (e.g., Didiereaceae, Crowley & Godfrey 2013). Furthermore, thorns are often produced well above the reach of tortoises and elephant birds, perhaps to deter arboreal lemur species that climb trees to feed on leaves in forest canopies.

Madagascar's unusually diverse vertebrate herbivore fauna complicates comparisons of plant defensive adaptations with other isolated islands. In particular, the presence of mammalian herbivores confounds tests for convergence in plant defensive adaptations against toothless browsers. Given that nearly all of these herbivores are extinct, matching putative defensive adaptations to particular types of herbivores may now be impossible.

Mascarene Islands

Approximately 1,000 km off the east coast of Madagascar lies the first in a series of oceanic islands extending in an arch stretching in a north-easterly direction across the Indian Ocean. Reunion (2,512 km^2) is the largest and closest island to Madagascar, followed by Mauritius approximately 175 km to the north-east (1,865 km^2), and Rodrigues, a further 500 km east (104 km^2). The Mascarene Islands were created by a 'hot spot' in the earth's crust, where magma escapes from the earth's mantle and then cools. As volcanic debris builds up, it breeches the ocean's surface to form islands, which in the case of the Mascarenes have moved slowly in a north-east direction on the African plate. Therefore, larger, younger islands are located farthest south-west, closest to the hot spot.

The three main Mascarene Islands have related floras that are derived from overseas dispersal, mainly from Madagascar and Africa. They also housed similar assemblages of vertebrate herbivores in their recent past (Cheke & Hume 2008). Giant tortoises occurred on all three main islands (*Cylindraspis* spp.). Dodos (*Raphus cucullatus*: Columbidae) occurred on Mauritius, and a related species, the Rodrigues solitaire (*Pezophaps solitaria*: Columbidae) occurred on Rodrigues Island. However, the extent to which they browsed on leaves is unclear, and their diets may have been dominated by fruits and bulbs, rather than leaves (Cheke & Hume 2008).

The incidence of leaf spinescence in the Mascarenes has yet to be quantified and compared with archipelagos. Similarly, the comparative incidence of divaricate branching has also yet to be established. However, leaf heteroblasty is common (Friedmann & Cadet 1976). Many species of woody plants that are endemic to the Mascarenes produce juvenile leaves that differ markedly in size, shape, and colour to their adult leaves.

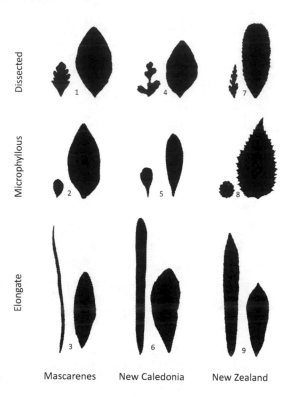

FIGURE 2.2 Silhouettes of species displaying leaf heteroblasty in the Mascarene Islands (left column), New Caledonia (middle column), and New Zealand (right column). Juvenile leaves are shown on the left and adult leaves are shown on the right. Leaf heteroblastic species in all three archipelagos fall into distinct morphological categories: dissected (top row), microphyllous (middle row), and elongate (bottom row). (1) *Quivisia heterophylla*: Meliaceae, (2) *Securinega durissima*: Phylanthaceae, (3) *Elaeodendron orientale*: Celastraceae, (4) *Codiaeum peltatum*: Euphorbiaceae, (5) *Atractocarpus rotundifolius*: Rubiaceae, (6) *Streblus pendulinus*: Moraceae, (7) *Elaeocarpus hookerianus*: Elaeocarpaceae, (8) *Hoheria sexstylosa*: Malvaceae, (9) *Elaeocarpus dentatus*: Elaeocarpaceae. The silhouettes are not drawn to scale.

The morphology of juvenile leaves fall into three distinct categories (Fig. 2.2). The juvenile leaves of some species are similar in shape to adult leaves, yet they are markedly smaller (i.e., 'microphyllous'). Other species produce heavily lobed, compound or palmate leaves at early ontogenetic stages (i.e., 'dissected'). However, the most common ontogenetic shift in leaf morphology is characterised by juvenile leaves that are markedly longer and thinner than adult leaves (i.e., 'elongate'). Interestingly, long, thin juvenile leaves are often coloured differently than adult leaves. While adult leaves tend to be more uniformly green, elongate juvenile leaves often produce brightly coloured mid-veins.

Hansen et al. (2003) tested whether conspicuously coloured, juvenile leaves have higher concentrations of chemical defences. In a thorough screen of many heteroblastic species from the Mascarenes, they found no evidence that juvenile leaves were better defended chemically than adult leaves. Therefore, the bright colouration associated with many juvenile leaves could not be interpreted as an aposematic warning of elevated chemical defence. Nevertheless, giant tortoises tend to avoid consuming juvenile leaves in favour of adult leaves.

Eskildsen et al. (2004) offered adult and juvenile leaves of a number of heteroblastic species, as well as a number of co-occurring homoblastic species, to giant Aldabran tortoises (*Aldabrachelys gigantea*: Testudinidae). They found that tortoises consistently avoided juvenile leaves. Yet, why tortoises avoid juvenile leaves remains unclear. It could be that microphyllous leaves provide a lower energetic return to consumers. Highly dissected leaves could complicate the process of acquiring search images for palatable plant species. The conspicuously coloured mid-veins of elongate, juvenile leaves may signal greater tensile strength. However, these hypotheses have yet to be tested and additional study is needed to pinpoint the characteristics of juvenile leaves that tortoises find less palatable.

Socotra

Like Madagascar, Socotra sits on an ancient granite block that was once part of eastern Gondwana. However, it lies north of Madagascar, 100 km east of the Horn of Africa. It became isolated from the Arabian Peninsula much later, approximately 18 Ma (Van Damme & Banfield 2011; Culek 2013). It currently supports 800 species of plants, 37% of which are endemic (Van Damme 2009).

Fossil or subfossil evidence for large vertebrate herbivores has yet to be found on Socotra. However, there is a long history of human occupation and exploitation of the island, and an ancient text written approximately two millennia ago describes Socotra as the home of tortoises. The text mentions two tortoise taxa in particular, one that occurred at higher elevations and another that occurred in lowland areas (see Farmer 2014).

Large thorns (<2 cm long) are rare in the flora of Socotra, which is striking given its close proximity to mainland Africa, where large thorns are commonplace (Farmer 2014). However, many plant species on Socotra produce smaller-sized prickles, thorns, or spines. Notable examples occur in several genera in the Acanthaceae (*Barleria*, *Blepharis*, *Justicia*) and Laminaceae (*Leucas*). Spinescent Socotran plants tend to be small in stature and often have a distinctive branching pattern whereby proximal leaves are protected by outward-pointing, spinescent stems (Fig. 2.3). Plants with similar morphological characteristics also occur in New Zealand and have been referred to as 'porcupine plants' (Burns 2016a). Quantitative comparisons between the incidence of different types of spinescence on Socotra and nearby

FIGURE 2.3 A spinescent shrub species from Socotra (*Neuracanthus aculeatus*: Acanthaceae) that produces dense networks of spines along the exterior of its canopy, which may have protected the leaves below from tortoise herbivory. (photo taken by Edward Farmer)

Africa have yet to be conducted, but are needed to test the insular spinescence hypothesis explicitly.

More distinctive spinescent structures are produced by other Socotran taxa. For example, *Tragia balfouriana*: Euphorbaceae produces poisonous, spinescent structures similar to the stinging nettle (*Urtica* spp.). Endemic species of *Hibiscus*: Malvaceae produce sharpened hairs that become dislodged from leaves when they are disturbed mechanically and rain down from the canopy on objects below (Farmer 2014). Heterophylly and heteroblasty are also commonplace. Most notably, frankincense (*Boswellia elongatata*: Burseraceae) produces long, narrow lanceolate leaves as juveniles, and then much larger, pinnately compound leaves as adults. This shift in morphology between juveniles and adults is similar to the elongate juvenile leaves produced by many heteroblastic species in the Macarenes, as well as New Zealand and New Caledonia (Fig. 2.2). However, the overall incidence of leaf heteroblasty in the Socotra flora is unclear, and whether the size and shape of juvenile leaves conform to the three types of morphological transitions present on other toothless-browser islands is unknown.

Hawai'i

The world's most isolated oceanic archipelago is located 3,500 km from the west coast of North America in the eastern Pacific Ocean. The Hawai'ian Islands are a hot spot archipelago similar to the Mascarene Islands. They consist of four main islands and numerous smaller islands that stretch in a north-westerly direction away from the hotspot (18° 55' 12.00" N, −155° 16' 12.00" W). Younger islands are larger and located closer to the hot spot than older islands (Hawai'i: 0.4 Ma, 10,433 km²; Maui: 0.8–1.3 Ma, 1,883 km²; O'ahu: 2.6–3.75 Ma, 1,545 km²; Kauai: ~ 5.1 Ma, 1,431 km²; Walker 1990; Neal & Trewick 2008).

No evidence for giant tortoises has ever been unearthed from Hawai'ian sediments. Instead, subfossils of two evolutionary lineages of avian herbivores have been found. Unfortunately, all but one failed to escape extinction following human arrival approximately 1,600 years ago. The first lineage is known as 'moa-nalo' and evolved from dabbling ducks (Anatidae). Several species of moa-nalo are currently recognised, and their subfossil remains have been found on all of the major islands except for Hawai'i (Slikas 2003). Moa-nalo were strongly herbivorous and may have specialised on ferns (James & Burney 1997).

The second lineage of avian herbivores was superficially similar to the moa-nalo, but evolved from geese rather than ducks. The largest species, the giant Hawai'i goose (*Branta rhuax:* Anatidae), was restricted to the youngest island (Hawai'i) and weighed approximately 8 kg. The nēnē-nui (*Branta hylobadistes*) was a slightly smaller species that inhabited many of the older islands in the archipelago (Paxinos et al. 2002). A third species, the nēnē (*Branta sandvicensis*) is the only endemic *Branta* species to escape extinction. Although capable of flight, the nēnē spends most of its time foraging on the ground, and, until recently, it occurred on all of Hawai'i's major islands.

Spinescence has long been argued to be uncommon in the Hawai'ian islands (Carlquist 1980). However, there are several notable exceptions. The Hawai'ian nightshade or popolo (*Solanum incompletum:* Solanaceae) is a distinctive-looking, endemic plant that produces remarkably large, conspicuous spines on both the upper and lower surfaces of its leaves, which appear bright red to human eyes.

Cyanea (Campanulaceae), the largest genus in the Hawai'ian Islands, contains many spinescent species. Of the 55 taxa considered by Givnish et al. (1994), 15 produce thorn-like prickles, a trait that evolved autochthonously on at least four occasions. Prickles are strongly heteroblastic in many of these species and are produced preferentially when plants are small, during early ontogenetic stages. Heteroblasty in spinescence is also frequently associated with heteroblastic shifts in leaf shape. Many spinescent species produce deeply lobed leaves at earlier ontogenetic stages, which is similar to the 'dissected' class of juvenile leaves in the Mascarenes.

FIGURE 2.4 Prickles produced by *Rubus hawaiensis* (Rosaceae). Photo taken by Philip Thomas

Carlquist (1980) used the Hawai'ian raspberry, or akala (*Rubus hawaiensis*: Rosaceae), as an example of the loss of herbivore defence on islands, as adult plants tend to lack structural defences, unlike its closest mainland ancestor, *Rubus spectabilis* (Morden et al. 2003). However, upon closer inspection, younger plants can be strongly spinescent (Fig. 2.4), which may have helped protect them against herbivory by giant ducks and geese. Some species in the genus *Rubus* use recurved prickles as aides for climbing other plants for structural support (e.g., *Rubus australis*), so spinescence in *R. hawaiensis* may have a functional significance that is unrelated to herbivory. However, Randell et al. (2004) showed that the prickles produced by *R. hawaiensis* are straight rather than recurved.

The Hawai'ian prickly poppy (*Argemone glauca*: Papaveraceae) is another notable spinesent plant species in Hawai'i. It produces prickles on leaves, stems, and fruits, in addition to latex when wounded. Barton (2014) showed that prickles were produced in similar densities throughout ontogeny. However, Hoan et al. (2014) showed that increased prickle densities could be induced by mechanical damage and that on younger plants the magnitude of prickle induction was higher.

Galapagos

The Galapagos is another 'hot spot' archipelago that has been volcanically active for over 20 Ma. Located 900 km off the Pacific coast of South America, islands in the Galapagos do not align in an arch like the Hawai'ian Islands or the Mascarenes. Instead, the 20 or so islands that make up the archipelago are distributed more contiguously.

Although devoid of large browsing birds, many islands in the Galapagos supported giant tortoises (*Chelonoidis* spp.), while others appear to have never housed tortoise herbivores. As a result, convergence in plant defence in the presence of toothless browsers, and the loss of defence on islands without vertebrate herbivores (i.e., defence displacement) can be tested among islands within the archipelago. This unique set of circumstances makes the Galapagos an especially interesting place to investigate repeated patterns in the evolution of plant defence. However, quantitative investigations of structural defences against vertebrate herbivores have yet to be conducted (*c.f.* Adsersen & Adsersen 1993).

There is a long history of speculation that the morphology of 'prickly-pear' *Opuntia* cacti varies among islands according to the distribution of herbivores (Nicholls 2014). There are six species of *Opuntia* in the Galapagos, all of which are a preferred food of giant tortoises. Stewart (1911) speculated that the morphology of *Opuntia* cacti covaried with the distribution of giant tortoises. On islands that supported tortoises in the recent past, juvenile *Opuntia* cacti tend to produce long spines that are rigid and sharp. However, once they grow above the reach of giant tortoises, their pad-like stems apparently produce soft, bristle-like spines. A very different pattern emerges on islands that failed to receive tortoise colonists. On herbivore-free islands, *Opuntia* species seem to be smaller statured or prostrate, and markedly less spinescent overall (Dawson 1966).

New Zealand

New Zealand is a continental archipelago comprised of two main islands and hundreds of smaller islands located approximately 1,500 km east of Australia. Like Madagascar and Socotra, it was previously connected to the supercontinent Gondwana. However, it began to rift away from Australia c. 83 Ma, becoming completely independent in its current latitudinal position c. 40 Ma. Although some components of New Zealand's flora and fauna may have been present at the time it split from Gondwana, much of New Zealand's flora is derived via overwater dispersal, mostly from Australia (see Gibbs 2006).

New Zealand was a land dominated by birds prior to the arrival of humans (Lee et al. 2010). In addition to flightless rails (e.g. takahe, *Porphyrio* spp.) and herbivorous geese (*Cnemiornis* spp.), New Zealand was home to an unparalleled diversity of giant, flightless ratites with strongly herbivorous diets. Eight species of 'moa' are currently recognised (Dinornithiformes), all of which went extinct soon after human arrival approximately 750 years ago (McCulloch & Cox 1992; Tennyson & Martinson 2006). Measured in generations of woody plants, this change in the ecology of New Zealand has occurred very recently, too recently for evolution to favour the loss of defensive adaptations geared specifically towards toothless browsers.

Chemical defences in the New Zealand flora are poorly understood. Pollock et al. (2007) conducted cafeteria-style experiments using ostriches (*S. camelus*) as a surrogate for extinct avian herbivores. They found that ostriches tend to avoid plants with high levels of phenolics. However, the effects of plant chemistry on consumption rates were overshadowed by the effects of leaf size and branching architecture.

Divaricate branching is an outstanding feature of the New Zealand flora. Although plants with this distinctive growth form occur elsewhere on the planet, it appears to be unusually common in New Zealand. Nearly 10% of all woody plant species native to New Zealand are divaricately branched and the growth form has evolved independently in 17 plant families, including both angiosperms and gymnosperms (Greenwood & Atkinson 1977; Dawson 1988). Although the incidence of divaricate branching seems unusually high in New Zealand, quantitative biogeographic comparisons of the incidence of divaricate branching have never been conducted.

A mechanistic explanation for the seemingly high incidence of divaricate branching in New Zealand is highly contentious. Divaricate branching could enhance physiological performance in specific environmental conditions, some of which may have been more prevalent in the recent past. For example, it could promote foliar frost tolerance, light capture, or prohibit photo-inhibition (e.g., Day 1998; Howell et al. 2002; Christian et al. 2006). However, empirical evidence for a link between divaricate branching and plant physiological performance is equivocal (Gamage & Jesson 2007).

There is a long history of speculation that divaricate branching evolved as a defence against avian browsers (Carlquist 1974; Greenwood & Atkinson 1977). Unfortunately, direct tests of this hypothesis are no longer possible because the putative selection agents are now extinct. However, Bond et al. (2004a) demonstrate that the shoots of several divaricately branched plant species are attached to stems more strongly than their non-divaricately branched relatives. They suggest that the increased tensile strength of stems, coupled with their low energetic yield given the small size and sparse spacing of leaves, would make it difficult for a toothless browser to make a living foraging on divaricate plants.

In food choice experiments using surrogate ratite browsers (emus, *D. novaehollandiae*), Bond et al. (2004a) found that divaricate plants tend to be avoided in favour of non-divaricate plants (see also Pollock et al. 2007). This provides strong, albeit indirect, support for the moa hypothesis. Although moa and emu could have foraged similarly, the avian browsers that inhabited New Zealand display a wide diversity of cranial and bill morphology and likely foraged in different ways (Attard et al. 2016). As a result, we will never know the extent to which emu are an accurate surrogate for moa in food choice experiments.

Divaricately branched tree species tend to be strongly hetero-blastic. Younger plants are divaricately branched and as they mature they abruptly begin to produce larger leaves, shorter internodes, and shallower branch angles. Transition heights between divaricately branched juveniles and adults typically occur approximately 3 m above the ground, which broadly corresponds to the height of the tallest known moa species (see Greenwood & Atkinson 1977; Atkinson & Greenwood 1989), and is therefore consistent with the defence displacement hypothesis. However, in this instance, plants are separated from herbivores vertically, rather than biogeographically.

In contrast to divaricate branching, spinescence is not a prominent feature of the New Zealand flora (Box 2.1). Only 12 native taxa produce prickles, thorns, or spines (Fig. 2.5). However, in eight of these species, spinescence is associated with leaves, which is proportionally higher than its incidence in Africa (Milton 1991).

Many spinescent plant species around the globe deploy prickles, thorns, and spines plastically. After being damaged mechanically, plants often increase the production of spinescent structures (Obeso 1997; Gómez and Zamora 2002; Göldel et al. 2016). This allows plants to deploy prickles, thorns, and spines strategically, at times and places where they may be more likely to be attacked by herbivores. Leaf spinescence in some New Zealand plants, which were subject to toothless rather than mammalian browsers, can also be induced by mechanical damage (Box 2.2).

Leaf heteroblasty is a notable feature of the New Zealand flora. A large number of woody plant species from diverse phylogenetic backgrounds produce differently shaped leaves through ontogeny. Some produce microphyllous juvenile leaves, often in conjunction with divaricate branching (e.g., *Carpodetus serratus*: Rousseaceae; *Pennantia corymbosa*: Pennantiaceae; *Sophora microphylla*: Fabaceae). Others produce long, thin juvenile leaves (e.g., *Elaeocarpus dentatus*: Elaeocarpaceae; *Knightia excelsa*: Proteaceae), or highly dissected and/or deeply lobed juvenile leaves (*Raukaua simplex*: Araliaceae). Categories of leaf shape at the juvenile phase in New Zealand

BOX 2.1   **Insular spinescence**

There are 12 currently recognised spinescent plant taxa in New
Zealand (Table B2.1, see also Burns 2016a). Some taxa, such as
Pseudopanax spp.: Araliaceae, likely evolved to become spinescent
autochthonously. However, other taxa (e.g., *Discaria toumatou*:
Rhamnaceae; *Eryngium vesiculosum*: Apiaceae; *Rubus* spp.: Rosaceae;
Urtica spp.: Urticaceae) also occur in Australia, and therefore could
have evolved spinescent structures prior to their arrival in New
Zealand.

Spinescence can arise developmentally from several types of plant
tissues, including leaves (spines), stems (thorns), and epidermal tissue
(prickles). Two-thirds of spinescent plant taxa in New Zealand produce
leaf spines. Given that birds lack teeth and cannot chew, leaf spines
may be particularly effective at deterring bird browsers, more so than
thorns or prickles located on branches.

Is the incidence of leaf spines unusually common in New Zealand?
Two-thirds of spinescent plant lineages in New Zealand deploy

Table B2.1 *Native species of spinescent plants in New Zealand
and type of spinescence*

Species	Type of spinescence
Aciphylla spp.	Leaf spines
Aristotelia fruticosa	Thorns
Carmichaelia odorata	Thorns
Discaria toumatou	Thorns
Eryngium vesiculosum	Leaf spines
Leptecophylla juniperina	Leaf spines
Melicytus alpinus	Thorns
Olearia ilicifolia	Leaf spines
Podocarpus spp.	Leaf spines
Pseudopanax spp.	Leaf spines
Rubus spp.	Prickles on leaves & stems
Urtica spp.	Poisonous spines on leaves & stems

* Genera containing multiple spinescent species

BOX 2.1 (cont.)

spinescence in close association with leaf tissues (Table B2.1). Milton (1991) provides comparable figures for spinescent plants in Southern Africa, where leaf spines were produced by approximately 23% of spinescent taxa, across a range of habitats, including forests, deserts, and grasslands. A binomial test for differences in the proportion of species producing leaf spines between regions indicates that leaf spinescence is unusually common in New Zealand ($p = 0.002$).

FIGURE 2.5 *Olearia ilicifolia*: Asteraceae, a spinescent plant species endemic to New Zealand.

plants are generally consistent with those found on the Mascarene Islands (microphyllous, elongate, dissected, Fig. 2.2).

Elaeocarpus hookerianus: Elaeocarpaceae provides a particularly striking example of leaf heteroblasty. Juvenile plants produce leaves with a remarkable range of shapes, from elliptical to linear with irregular and asymmetric lobbing (Fadzly & Burns 2010, see also Day & Gould 1997; Day et al. 1997). Another striking feature of *E. hookerianus* is that its juvenile leaves have high concentrations of

BOX 2.2   **Induced spinescence**

Induced defence is common in spinescent plants that are exposed to mammalian herbivory (Barton 2016; Coverdale et al. 2018). When damaged by herbivores, spinescent plants inhabiting continents often increase investment into prickles, thorns, and spines. To test whether plant species that are endemic to islands with toothless herbivore islands might also display induced defences, the size of lateral leaf spines was compared between damaged and undamaged saplings of a spinescent plant species from New Zealand.

Pseudopanax crassifolious: Araliaceae is a common tree species in New Zealand. It commonly grows alongside hiking trails, which are periodically cleared of overhanging vegetation. This sometimes removes the apical meristems of *P. crassifolious* saplings, many of which survive to produce new leaves.

To test for the induction of leaf spines following damage, I measured the length (mm) of the largest leaf spine on the youngest mature leaf of 25 damaged plants growing alongside trails traversing Nelson Lakes National Park, New Zealand (41°48' S, 172°50'E). Spine size in the youngest mature leaf on the closest undamaged plant located nearby was also measured (n = 25).

The size of lateral leaf spines covaries with several aspects of leaf morphology. Lateral leaf spines increase in size with both leaf area and the stature of juvenile plants. To control for these potentially confounding effects, leaf width (mm) and plant height (cm) were also measured.

A general linear model was then used to test for differences in spine size between damaged and undamaged plants. Relative spine size (spine length·leaf width^{-1}) was used as the dependent variable. Damage was considered a fixed factor with two levels (i.e. damaged vs undamaged) and plant height was used as a covariate. Relative spine size was log transformed to conform to assumptions prior to analyses.

Results showed that spine size increased with plant height (F = 4.277, p = 0.044), at similar rates (i.e., similar slopes) for both damaged and undamaged plants (F = 0.736, p = 0.396). However, damaged plants had a higher intercept than undamaged plants (F = 22.973, p < 0.001),

BOX 2.2 **(cont.)**

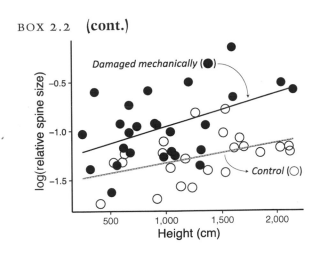

FIGURE B2.2 Induced defence in lancewood (*Pseudopanax crassifolious*: Araliaceae). Relative spine size (spine length·leaf width^{-1}, y-axis, log transformed) is plotted against plant height (x-axis) in damaged (black circles, solid line) and undamaged (white circles, dashed line) plants.

indicating that damaged plants induced larger leaf spines following damage (Fig. B2.2). All analyses were performed in the R environment (R Core Team 2013).

accessory pigments (e.g., anthocyanins) leading to leaf hues ranging from green to mottled brown or even black to the human eye. When quantified from the perspective of birds, spectrographic analysis indicates that juvenile leaves would have been difficult to distinguish visually against a background of leaf litter (Fadzley & Burns 2010). On the other hand, adult leaves are regularly oblong, appear green to human observers, and standout clearly against litter.

New Caledonia
Several thousand kilometres north of New Zealand lies an isolated island with a similar geologic history. New Caledonia separated from Australia 65 Ma, moving in a north-easterly direction into the tropical

Pacific. It has an exceedingly diverse flora that was subject to several types of large herbivores prehistorically. In addition to herbivorous tortoises (Hawkins et al. 2016), New Caledonia was also home to at least one species of giant, flightless, browsing bird (*Sylviornis neocaledoniae*: Sylviornithidae; Worthy et al. 2016).

Very little is known about how New Caledonian plants may have defended themselves against toothless herbivores. However, leaf heteroblasty is commonplace (Burns & Dawson 2006). A diverse range of species abruptly change leaf morphology during development. The juvenile leaves can also be placed in the three morphological categories present in both New Zealand and in the Mascarene Islands, i.e., dissected, elongate, and microphyllous (Fig. 2.2). Leaf heteroblasty therefore appears to be a convergent phenomenon on several toothless browser islands distributed across the globe. However, a formal quantitative investigation for congruence in leaf heteroblasty among the Mascarenes, New Caledonia, and New Zealand and other toothless browser islands have yet to be conducted.

Islands without Vertebrate Herbivores

Chatham Islands

Approximately 4 Ma, the Chatham Islands uplifted above the Pacific Ocean, 650 km east of New Zealand. The present-day archipelago, which is comprised of one large island (Chatham Island) and several smaller islets, has never been connected to a larger landmass and its flora is derived entirely by over-water dispersal from nearby landmasses, mostly from New Zealand. No evidence has ever been found for the existence of flightless vertebrate herbivores on the Chatham Islands prehistorically. The defence displacement hypothesis therefore makes clear predictions for the Chatham Islands. Selection should favour the loss of defensive adaptations against avian herbivores in species that evolved from ancestors in New Zealand, where they were exposed to avian herbivores.

Greenwood and Atkinson (1977) compared the number of divaricately branched species inhabiting coastal regions of New

Zealand to that found on the Chatham Islands. They found that divaricate plants were less prevalent on the Chatham Islands, as well as several other oceanic islands that flank New Zealand, suggesting a repeated pattern in the loss of divaricate branching in the absence of avian herbivores. However, an alternative explanation could be that the number of divaricate species on herbivore-free islands results from selective immigration, if divaricate plants are relatively poor over-water dispersers.

To explore Greenwood and Atkinson's (1977) conclusions further, Kavanagh (2015) compared leaf size and branching architecture in Chatham Island plant species to their closest relatives in New Zealand. He found that Chatham Island plants consistently produced larger leaves, shorter internodes, and narrower branch angles than their ancestors in New Zealand. Consequently, in the absence of selection from toothless browsers, Chatham Island plants appear to have evolved the loss of divaricate branching (Fig. 2.6; Box 2.3).

Spinescence is rare among plants on the Chatham Islands. However, several endemic species evolved from spinescent ancestors in New Zealand. For example, two species of speargrass (*Aciphylla dieffenbachia* and *A. traversii*: Apiaceae) occur on the Chatham Islands, both of which are derived from spinescent ancestors in New Zealand. Burns (2016a) compared leaf compression strength and the size of terminal leaf spines between Chatham Island and New Zealand taxa and found that *A. dieffenbachia* produces less-rigid leaves with smaller terminal spines. Box 2.4 illustrates similar results for sister species of *Leptecophylla*: Ericaceae.

While leaf shape heteroblasty is common in New Zealand, it is rare on the Chatham Islands (Burns & Dawson 2009). Two tree species that are endemic to the Chatham Islands produce heteroblastic leaves (*Dracophyllum arboretum*: Ericaceae; *Coprosma chathamica*: Rubiaceae). However, differences between the morphology of juvenile and adult leaves in both of these species differ markedly from the categories of morphological change found in the Mascarene Islands, New Caledonia, and New Zealand (Burns & Dawson 2009). All three

FIGURE 2.6 Photographs of three shrub species endemic to the Chatham Islands, which lacked vertebrate herbivores prehistorically: (a) *Myrsine chathamica*: Primulaceae; (c) *Corokia macrocarpa*: Argophyllaceae; (e) *Melicytus chathamica*: Violaceae

(e)

(f)

FIGURE 2.6 (*cont.*) Alongside are their closest ancestors in New Zealand, where they were subject to avian herbivory throughout their evolutionary history: (b) *Myrsine divaricata*; (d) *Corokia cotoneaster*; (f) *Melicytus alpinus*. All three Chatham Island endemics illustrate an evolutionary loss of divaricate branching (Kavanagh 2015), including a decline in leaf tensile strength (Box 2.3).

Chatham Island species produce juvenile leaves that are similarly shaped, but much larger than adults. Therefore, in addition to the loss of divaricate branching, spinescence, and leaf colouration, selection also appears to have favoured the loss of leaf heteroblasty on the Chatham Islands, or at least the type of heteroblasty found on other isolated islands.

Iceland

Three hundred kilometres east of Greenland lies a large, volcanic island that was devoid of large vertebrate herbivores prior to the arrival of humans (Darlington 1957; Bryant et al. 1989). Plant communities growing at similar latitudes in Asia, Europe, and North America are subject to a wide assortment of vertebrate herbivores, including

BOX 2.3   **Loss of tensile strength in divaricate shrubs**

The tensile strength of leaves produced by divaricate plants in New Zealand tends to be higher than their non-divaricate relatives (Bond et al. 2004a). This suggests that avian herbivores that forage for leaves individually using a plucking motion would have to pull harder to remove leaves of divaricate species. If this hypothesis is correct, then increased leaf tensile strength should be lost in populations on islands that lacked avian herbivores.

The Chatham Islands are a geologically young archipelago that acquired its flora entirely by over-water dispersal, mostly from New Zealand. To test for the loss of leaf tensile strength in divaricate plants in the absence of avian herbivory, the force required to remove the leaves of three divaricately branched species in New Zealand were compared to their sister species on the Chatham Islands, which lacked avian herbivores. Heenan et al. (2010) illustrated that the closest ancestor of three Chatham Island species, *Corokia macrocarpa*: Argophyllaceae, *Melicytus chathamicus*: Violaceae, and *Myrsine chathamica*: Primulaceae are the divaricately branched New Zealand endemics *Corokia cotoneaster*, *Melicytus alpinus*, and *Myrsine divaricata*, respectively. To compare their tensile strength, a single leaf from 17–30 plants of each species was measured using a force metre. Leaf size (leaf length·leaf width) was also measured as a course estimate of the energetic return to potential consumers. Measurements were made in in Nelson Lakes National Park, New Zealand, and several forest reserves on Chatham Island (see Cox & Burns 2017 for site descriptions). Leaf tensile strength (Newtons, N) was then divided by leaf area (cm^2) and compared between species pairs using separate linear models.

Results showed that relative leaf tensile strength tended to be higher in New Zealand (Fig. B2.3). Ratios between tensile strength and leaf area differed in all three comparisons (*C. cotoneaster* [n = 20] vs *C. macrocarpa* [n = 30], t = 2.148, p = 0.037; *M. alpinus* [n = 17] vs *M. chathamicus* [n = 30], t = 10.895, p < 0.001; *M. divaricata* [n = 30] vs *M. chathamica* [n = 30], t = 4.813, p < 0.001). Therefore, relative to their size, the leaves of New Zealand species were more difficult to remove

BOX 2.3   **(cont.)**

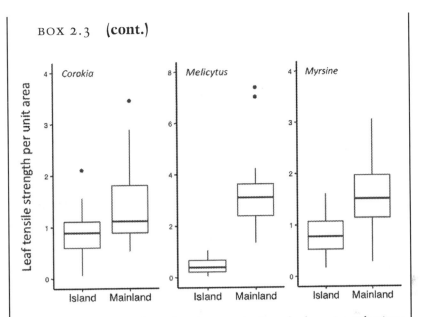

FIGURE B2.3 Relative foraging costs in three shrub species endemic to the Chatham Islands (labelled 'Island' on the x-axis), which evolved from divaricate ancestors in New Zealand (labelled 'Mainland' on the y-axis). The ratio between the force required to remove a leaf from its stem (N) and its area (cm^2) is shown on the y-axes. Relative foraging costs were statistically higher in New Zealand for all three taxonomic pairings.

than Chatham Island species. This suggests that selection has favoured higher foraging costs in New Zealand, where plants were exposed to avian herbivores (i.e., defence displacement). The data were courtesy of Bart Te Manihera Cox.

deer and caribou (Cervidae). Prehistorically, the diversity of large herbivores in these regions was even higher (Martin & Klein 1989).

Chemical defences are common in high latitude forests and scrublands, with smaller, younger plants typically being better defended than larger, older plants (Swihart & Bryant 2001). Bryant et al. (1989) compared chemical defences between Icelandic birch

BOX 2.4 Loss of spinescence in *Leptecophylla*

Leptecophylla juniperina: Ericaceae is a widespread shrub species in New Zealand that produces small, needle-like leaves, which could have been difficult for avian browsers to swallow whole.
Leptecophylla robusta, is a closely related species that is endemic to the Chatham Islands, which lacked large, avian herbivores. Molecular evidence indicates that *L. robusta* recently diverged from *L. juniperina* after dispersing to the Chatham Islands (Heenan et al. 2010).

To test whether *L. robusta* has evolved to become less spinescent than *L. juniperina* in the absence of avian herbivory in the Chatham Islands, leaf rigidity (i.e., compression strength) and the size of terminal leaf spines were compared between taxa. Measurements were made on a single leaf from 25 individuals of *L. robusta*, in several forest reserves on the Chatham Islands (see Cox & Burns 2017 for site descriptions) as well as 25 individuals of *L. juniperina* from Nelson Lake National Park, South Island, New Zealand. Leaf compression strength was measured in Newtons (N) as the amount of force required to buckle leaves when compressed longitudinally. The length of the terminal leaf spine was measured to the nearest 0.01 mm using a dissecting microscope. Both leaf compression strength and the size of terminal leaf spines covary passively with leaf size, so both variables were scaled by leaf area (leaf length·leaf width) prior to analyses. To test whether *L. juniperina* produces sharper, more-rigid leaves than *L. robusta*, separate linear models were conducted on each variable. Both variables were log transformed prior to analyses to conform to assumptions.

Results showed that spinescence was higher in New Zealand (Fig. B2.4). *L. juniperina* had longer terminal spines ($t = 8.423$, $p < 0.001$) and stiffer leaves ($t = 11.320$, $p < 0.001$) than *L. robusta*. Selection therefore seems to have favoured the evolution of softer leaves with smaller terminal spines in the absence of vertebrate herbivores (i.e., defence displacement). The data were courtesy of Bart Te Manihera Cox.

BOX 2.4 (cont.)

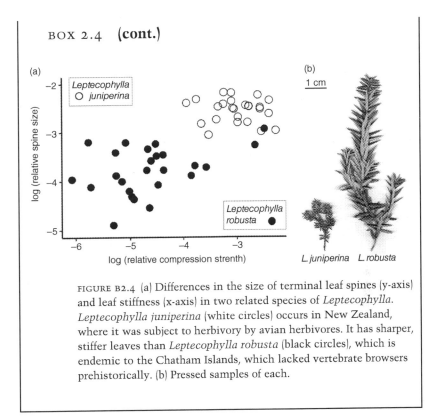

FIGURE B2.4 (a) Differences in the size of terminal leaf spines (y-axis) and leaf stiffness (x-axis) in two related species of *Leptecophylla*. *Leptecophylla juniperina* (white circles) occurs in New Zealand, where it was subject to herbivory by avian herbivores. It has sharper, stiffer leaves than *Leptecophylla robusta* (black circles), which is endemic to the Chatham Islands, which lacked vertebrate browsers prehistorically. (b) Pressed samples of each.

(*Betula pubescens*: Betulaceae) and a handful of related birch species from Finland, Siberia, and Alaska and found that young plants from Iceland had lower concentrations of internode resins and triterpenes than similar-statured plants sourced on continents. In a series of food choice experiments, they also offered island and mainland birch plants to snowshoe hares (*Lepus americanus*: Leporidae), a common arctic herbivore in Europe, Asia, and North America. Results showed that hares consistently preferred less-defended birch seedlings. Results from Bryant et al.'s (1989) thorough study therefore provide compelling evidence that selection has shaped the chemical signatures of island plants and they are now more susceptible to vertebrate herbivory.

Haida Gwaii

Previously known as the Queen Charlotte Islands, the Haida Gwaii archipelago is located in the Pacific Ocean off the north-west coast of North America. They sit at a similar latitude to Iceland, but have a very different geologic history. Haida Gwaii was connected to what is now northern Canada during the last glacial–interglacial period. However, it failed to acquire a diverse vertebrate fauna. A single subspecies of caribou (*Rangifer tarandus dawsoni*: Cervidae) managed to colonise the islands naturally. However, it was a relatively unimportant herbivore, as it was restricted to the largest island in the archipelago, Graham Island. Other islands in the archipelago have experienced between 15,000 and 30,000 years of isolation from vertebrate herbivores.

The region supports conifer forest dominated by a handful of tree species. Vourc'h et al. (2001) compared the chemical signatures of western red cedar (*Thuja plicata*: Cupressaceae) from Haida Gwaii to those on the mainland. They found that island plants had lower concentrations of terpenes and were therefore more poorly defended than mainland plants. When subjected to browsing by black-tailed deer (*Odocoileus hemionus sitkensis*: Cervidae), island plants were attacked more frequently, similar to Icelandic birch.

Lord Howe Island

Lord Howe Island is a small volcanic island located 600 km east of Australia. It has never been connected to a larger landmass and was devoid of vertebrate herbivores prior to human arrival. However, the bones of a distinctive, horned tortoise (*Meiolania platyceps*: Meiolaniidae) have been found in sediments dating to the Pleistocene (Gaffney 1996). The island was uninhabited prehistorically, so, unlike other island megafauna, it did not meet its demise at the hands of human hunters or their mammalian castaways. Instead, it went extinct thousands of years ago for unknown reasons and the island's flora has evolved in the absence of vertebrate herbivores ever since. Morphologically, the tortoise had a strongly domed shell and likely could not reach any more than 0.5 m above the ground (Gaffney 1996). Therefore, selection for plant traits aimed at deterring toothless herbivores would have been restricted to low-growing plants in the forest understory.

There are three spinescent plant taxa on Lord Howe Island, all of which are only slightly differentiated taxonomically from populations in Australia. *Smilax australis*: Smilacaceae is a liana that produces rigid prickles along its stems. *Alyxia ruscifolia*: Apocynaceae is a shrub that produces spade-shaped leaves that end in a sharp terminal spine, and *Drypetes deplanchei*: Putranjivaceae is a rainforest tree, whose seedlings produce holly-shaped leaves with sharp spines along their margins.

Burns (2013a; 2016b) compared the production of spinescent structures between island and mainland populations in all three taxa. In mainland populations of *S. australis*, smaller, younger stems had greater prickle densities than larger, older stems. However, on Lord Howe Island, plants were basically unarmed. Somewhat different results were obtained for *A. ruscifolia* and *D. deplanchei*. In these species, younger plants produced more spinescent leaves than older plants, both on the island and on the Australian mainland. However, spinescence was higher in Australia and mainland plants continued to produce spinescent leaves later into ontogeny than island plants.

California Islands

Situated less than 30 km off the coast of southern California lies an archipelago of 12 islands that are collectively known as the California Islands. Although they are weakly isolated in space, there is no evidence that they were connected to the mainland during the Pleistocene. Instead, they appear to have been isolated from North America for many more millennia (Johnson 1978). During this time, they have acquired a very distinctive fauna that is characterised by a lack of large herbivores, with one exception. Archaeological evidence indicates that it was once home to a dwarfed proboscidean, *Mammuthus exilis*: Elephantidae. Previous work suggested that its diet was comprised of grasses (Schoenherr et al. 2003). However, more recent research suggests that it may have been a browser, with a diet focused on leaves and twigs (Semprebon et al. 2016). Standing approximately 1.75 m tall, it was substantially smaller than its closest mainland relative, the Columbian mammoth (*Mammuthus columbi*), which stood well over 4 m tall (Agenbroad 2010). Therefore, it is uncertain whether trees and shrubs inhabiting the California Islands were completely free from the effects of vertebrate browsers, or whether they were subject to browsing by a single species of smaller-statured herbivore.

In a pioneering study of the loss of defence in island plants, Bowen and Van Vuren (1997) compared structural and chemical defences between populations of six species of trees and shrubs inhabiting Santa Cruz Island and the adjacent mainland. Spinescence was reduced in five out of six study species, while chemical defences showed weaker and less consistent changes. When fed to sheep, island plants were consumed in greater quantities than mainland plants, indicating that island plants were more poorly adapted to vertebrate browsers. Salladay (unpublished material, Herbivory Defense of Island Plants and their Mainland Relatives, Berkeley, CA) repeated Bowen and Van Vuren's (1997) study with nine species from a different island, Santa Catalina, and obtained similar results.

Burns (2013a) compared heteroblastic shifts in the production of leaf spines between plants on Santa Cruz Island and the adjacent mainland. Results showed that spine densities were highest in younger plants and declined as plants matured. However, spine densities declined more rapidly in island plants. Burns (2013a) interpreted this as an early, incomplete stage in the evolutionary loss of leaf spines. However, in light of recent paleontological evidence (Semprebon et al. 2016), an alternative explanation could be that selection has favoured a shift towards earlier production of adult leaves in a manner consistent with the reach height of the dwarfed mammoths.

Islands without Plant Protectors

Hawai'ian Islands

The Hawai'ian Islands lack native ants, so plants colonising the archipelago with extra-floral nectaries are unable to attract them as plant protection mutualists. Keeler (1985) showed that three native species (*Passiflora foetida*: Passifloraceae; *Ipomea indica*: Convovulceae; *Pteridium aquilinum*: Polypodiaceae), which produce extra-floral nectaries elsewhere in their range, do not produce extra-floral nectaries in Hawai'i. Therefore, in the absence of ants, they appear to have lost their extra-floral nectaries. Keeler (1985) also compared the incidence of extra-floral nectaries among thousands of introduced, native, and endemic species in the Hawai'ian flora. Results showed that fewer endemic species produced extra-floral nectaries than native species, and that fewer native species produced extra-floral nectaries than introduced species.

Ogasawara Islands

The Ogasawara Islands (formally known as the Bonin Islands) are located 1,000 km south of Japan in the western Pacific. They were created approximately 45 Ma when the Pacific plate began to subduct under the Philippine Sea plate. The resulting volcanic activity created an archipelago of around 30 islands, which received their floras via over-water dispersal, mainly from Japan.

The archipelago has a depauperate ant fauna (Sugiura et al. 2006). However, it was colonised by *Hibiscus tiliaceus*: Malvaceae, which produces extra-floral nectaries across its wide distribution spanning the western Pacific and Indian Oceans. A closely related species, *Hibiscus glaber*, is endemic to the archipelago and likely evolved from a common ancestor with *H. tiliaceus* in the recent past. It lacks extra-floral nectaries, suggesting selection has favoured the loss of extra-floral nectaries in the absence of mutualistic partners.

Interestingly, many species of ants have colonised the islands following the arrival of humans and some visit the extra-floral nectaries of *H. tiliaceus*. As a result, *H. tiliaceus* is rarely attacked by *Rehimena variegata* (Lepidoptera: Pyralidae), an endemic moth species that feeds on flowers. On the other hand, ants rarely visit *H. glaber*, presumably because they no longer provide energetic reward in the form of extra-floral nectar, and this endemic species is attacked frequently by *R. variegata*.

The Antilles

Cecropia (Urticaceae) is a genus of trees distributed throughout much of the neotropics and the islands in the Caribbean. A distinctive feature of most species in the genus is their mutualistic association with ants, most frequently in the genus *Azteca*: Formicidae. Most species of *Cecropia* produce Müllerian bodies at the base of leaf petioles, which are consumed by ants, who in turn vigorously defend plants against herbivores, as well as structurally parasitic plants (i.e., vines and lianas). However, islands in the Caribbean typically have less diverse herbivore communities, as well as fewer species of vines and lianas. Consistent with reductions in herbivore densities on islands, populations of *Cecropia* on Puerto Rico, Grenada, St Vincent, St Lucia, Barbados, Martinique, Dominica, and Guadalupe rarely produced Müllerian bodies (Janzen 1973; Rickson 1977). However, Trinidad and Tobago, two larger islands located closer to South America that house more diverse ant communities, support *Cecropia* plants that produce copious quantities of Müllerian bodies.

CONCLUSIONS

Overall results provide mixed support for syndrome predictions. On one hand, islands devoid of vertebrate herbivores repeatedly show evidence for defence displacement. Plant traits linked to mammalian herbivores are typically lost on islands devoid of vertebrate herbivores. Coniferous trees on Iceland and Haida Gwaii invest less in chemical defences than their mainland counterparts. Spinescent plants in Australia and California invest less in prickles, thorns, and spines after they colonise offshore islands. Plant traits that are thought to be effective at deterring toothless browsers also show evidence of defence displacement. Plants that colonised the Chatham Islands from New Zealand have repeatedly lost leaf spinescence, divaricate branching, and heteroblasty (including both leaf shape variation as well as apparently cryptic and aposematic leaf colouration).

Evidence for convergence in putative defensive adaptations against avian and reptilian herbivores on toothless browser islands is more equivocal. Contrary to some earlier accounts, spinescence is not absent from the floras of Hawai'i and New Zealand. It is also a conspicuous feature of the floras of the Galapagos and Socotra. Spinescence on these islands tends to be associated with leaves (i.e., spines) rather than stems (i.e., thorns), which may have been particularly effective at deterring toothless herbivores that would have had to swallow them whole. Divaricate branching is common in New Zealand and Madagascar, but it is curiously absent from Hawai'i. However, a global-scale analysis of the incidence of heteroblasty has yet to be conducted, so whether it has evolved convergently on islands with toothless browsers is untested. Evolutionary convergence in leaf heterobasty is perhaps the most consistent characteristic of islands that once housed toothless browsers, being similarly common on the Mascarene Islands, New Zealand, and New Caledonia.

Comparatively fewer studies have investigated changes in plant protection mutualisms on islands. However, all available information points to the consistent loss of extra-floral nectaries and Müllerian bodies on islands where plant protectors (e.g., ants) are absent. Therefore, on islands that lack plant protectors, selection for plant protection mutualisms appears to be relaxed and structures facilitating plant protection mutualisms are repeatedly lost evolutionarily. However, few studies have tested for the loss of plant protection mutualisms to date, despite much interest in the phenomenon in general.

Although great advances have been made towards a better understanding of plant defence on islands, many important questions remain. One obvious, overarching deficiency is that nearly everything we know comes from islands in the Indian or Pacific Oceans, and in particular New Zealand and its surrounding islands. Work in other archipelagos (e.g., Macaronesia) is needed to determine whether defence displacement occurs consistently across the globe. Another serious obstacle is the changes to herbivore communities brought about by humans. Given that many island mega-herbivores are now extinct, direct experimental tests of how plant traits affect the behaviour of toothless herbivores are often impossible. However, future study of plant defence in the Galapagos could be particularly informative, as it provides an unparalleled opportunity to explore variation in defensive adaptations among islands relative to one of the few toothless herbivores to escape extinction.

3 Differences in Dispersal

When viewed without an evolutionary perspective, insular flightlessness seems strange. How did flightless birds like the takahe get to isolated islands? Darwin was often pressured to answer this question. He responded with an analogy:

> As with shipwrecked mariners near a coast, it would have been better for the good swimmers if they had been able to swim still further, whereas it would have been better for the bad swimmers if they had not been able to swim at all and had stuck to the wreck.
>
> Darwin (1859, p. 177, cited in Lomolino 2010)

So what exactly is Darwin saying? In my opinion, he seemed to think there were two dispersal strategies for island organisms. On the one hand, if island organisms had exceptional powers of dispersal, they could avoid being lost at sea by dispersing back to the mainland. In other words, selection could favour long-distance dispersal. On the other hand, individuals could also avoid being lost at sea by staying put, or, in other words, selection could favour the loss of dispersal potential. This second strategy is often cited as a component of the island syndrome in animals. This chapter tests whether it is a repeated pattern in the evolution of island plants. It also explores whether the loss of dispersibility could arise as a passive by-product of selection for large seeds.

COSTS & BENEFITS OF DISPERSAL

The environment beneath a parent plant is not ideal for its offspring (Herrera 2002a). The most obvious disadvantage for seeds that fail to disperse is that they must compete with their parents for light, water, and nutrients. Avoiding parent–offspring conflict is therefore one obvious benefit of dispersal. The further seeds disperse away from their mother, the fewer siblings they are likely to encounter as well. So avoiding sibling competition is a second advantage of dispersal. However, there are additional advantages of dispersal.

Many plants are adapted to particular stages of succession that take place following disturbance. For these species, dispersal is a key feature of their life history because only by dispersing through space can they find sites undergoing particular stages of succession that they require in order to survive (Brokaw 1985; Lusk & Laughlin 2017). For example, shade-intolerant forest plants must disperse to recently disturbed tree-fall gaps in order to complete their life cycle.

A final benefit of dispersal is escape from species-specific seed predators and pathogens. Seed dispersal kernels are usually characterised by high seed densities close to parent plants and declining seed densities farther away. Therefore, density-dependent mortality resulting from specialist predators and pathogens is highest beneath parents, generating selection for increased dispersal distances to avoid succumbing to predatory and pathogenic attack (Janzen 1970; Connell 1971; Hyatt et al. 2003).

Despite its many advantages, dispersal can also be costly (Bonte et al. 2012). Perhaps the most obvious cost of dispersal is the energy required to produce accessory structures that help carry seeds to new locales. Fleshy fruit pulp, fluffy plumes, and hooked barbs often require substantial amounts of energy to produce. A second cost of dispersal is the possibility that seeds will arrive in locales that are not suitable for establishment. This cost is particularly relevant for islands plants. Given their close proximity to the ocean, island plants may often be swept out to sea to their deaths. This 'sea-swept' cost of dispersal forms the foundations to Darwin's thinking about the evolution of dispersal potential in island organisms.

DISPERSAL MECHANISMS

While conducting field work in south-western Australia, Carlquist (1976) stumbled across a strange species of dioecious plant that had yet to be described. At first, he could only locate male plants, but after more meticulous searching he solved the mystery of the missing females. Oddly, female flowers were produced underground. As a result, fruits and seeds were also located beneath the surface of the soil. *Alexgeorgea*: Restionaceae, Carlquist's name for the new genus, provides an excellent example of an unusual dispersal mode known as *geocarpy*, or the production of underground diaspores. Producing seeds underground presents an obvious limitation to dispersal distances, so island floras are typically devoid of geocarpic plant species.

Seed dispersal via wind is a common dispersal mechanism in plants. *Anemochory* has evolved repeatedly in plants and selection has modified a variety of plant tissues to help get seeds airborne. Some wind-dispersed plants produce winged 'samaras', or flattened invaginations of the ovary wall. These can be simple, sail-like structures (e.g., Nothofagaceae), or they can be scalloped into wings that auto-rotate (e.g., Aceraceae, Pinaceae, Proteaceae). Other wind-dispersed species produce fluffy plumes. Composites (Asteraceae) produce plumes that are derived from sepals, while in other taxa plumes are derived from different flower parts (e.g., *Epilobium* spp.: Onagraceae). Regardless of their anatomical origins, all wind dispersal aides reduce the terminal velocity of seeds, and the slower a diaspore falls to the ground, the more likely it is to intercept wind currents that could carry it aloft. However, when viewed from an insular perspective, the longer it takes for a wind-dispersed seed to settle on the ground, the more likely it is to be carried off an island and out to sea.

Similar to plants that release their seeds into air currents, coastal plants often release their seeds into the ocean to be dispersed by sea currents. Water-dispersed, or *hydrochorous*, fruits can achieve their buoyancy in several different ways. Some water-dispersed species produce seeds with pockets of air positioned alongside the embryo (e.g., *Hibiscus* spp.: Malvaceae). Other species produce spongy or fibrous tissues in their fruits that retain air and help them to float (e.g., *Cocos nucifera*: Arecaceae). The chemical composition of seed coats (e.g., waxy, oily, or mucilaginous testa) can also promote buoyancy. All of these attributes can potentially evolve independently of seed size and influence the length of time seeds can float on the surface of the sea.

Other plants employ the services of animals to disperse their seeds. Many plant species around the globe disperse their seeds mutualistically by surrounding their seeds with pulp to entice animals to swallow them whole. Once consumed, animals subsequently act as seeds vectors (i.e., *endozoochory*), depositing them in new locales sometime later after passing through the gut. Other plants enlist the services of animals as dispersers without their mutualistic consent. Adhesion dispersal is achieved by fruits with hooks, barbs, or sticky secretions that attach seeds to feathers, fur, or scales. These *epizoochorous* fruits arrive in new locations after they become dislodged (Sorsensen 1986).

Curiously, some plants produce diaspores that lack obvious dispersal aides altogether. Without the help of specialised structures such as wings or plumes, gravity-dispersed, or *barochorous*, plant species, are dispersed passively. When gravity-dispersed diospores are large, it severely constrains their dispersal distances, and seeds usually come to rest close to parent plants. When they are small, as is the case of orchids or bryophytes, they can be transported vast distances by air currents (see Wolf et al. 2001). Therefore, the size of gravity-dispersed diaspores directly affects their dispersal distances.

SYNDROME PREDICTIONS

Most fruits have two basic elements: seeds and structures that facilitate their dispersal. Fluffy plumes lift wind-dispersed seeds into air currents. Air bladders help water-dispersed seeds float on the surface of the sea. Hooked barbs attach animal-dispersed seeds to the feathers or fur of animals. The most obvious way to reduce the dispersal capacity of fruits is to reduce the functionality of dispersal aides. Smaller plumes, air bladders, and hooked barbs should lead to shorter seed dispersal distances, all else being equal. For the rest of this chapter, reductions in dispersal potential resulting from changes in dispersal aides *per se* will be referred to as the *loss of dispersibility hypothesis* (Fig. 3.1).

The size of dispersal aides and the seeds they transport typically covary with one another. Larger wind-dispersed seeds need bigger plumes

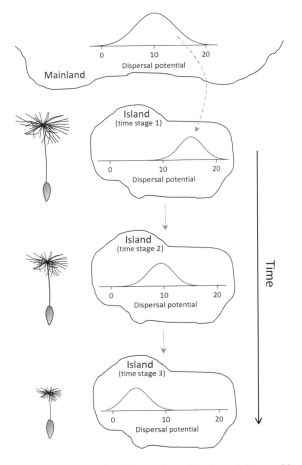

FIGURE 3.1 A graphical depiction of the loss of dispersal hypothesis. Three discrete time increments are illustrated. At time stage 1, islands are colonised from the mainland by individuals with good powers of dispersal. Because individuals with relatively good powers of dispersal are more likely to be lost at sea, and thus selected against, poor dispersibility should then evolve in the newly founded island population. Progressive loss of dispersibility is illustrated in time stages 2 and 3. Importantly, the evolutionary loss of dispersibility results specifically from reductions in the size and functionality of dispersal aides (i.e., a feathered plume). Adapted from Cody and McC. Overton (1996)

to be carried in the air. Larger hooks are necessary for larger seeds to attach to animals. Bigger air bladders are needed to float bigger seeds. As a result, differences in seed dispersal distances can arise not just from adjustments in the size of dispersal aides, but also from changes in seed size.

Many important aspects of the life history of plants can select for increased seed size. For example, larger seeds often give rise to larger, more competitive seedlings. Seedlings emerging from bigger seeds also tend to have higher survivorship and are more tolerant of environmental perturbations (Leishman & Westoby 1994; Harms & Dalling 1997; Geritz et al. 1999; Leishman et al. 2000; Coomes & Grubb 2003; Moles & Westoby 2004; Lönnberg & Eriksson 2013; Rubio de Casas et al. 2017). On the other hand, larger seeds are obviously more costly to produce.

The potential for evolutionary changes in seed size to effect dispersibility indirectly was not lost on Carlquist (1966b): "Increase in fruit size as a way of adaptation to shady forest conditions by virtue of a larger food storage on which seedlings can draw seems an operative factor in loss of dispersabilty" (p. 46). This alternative pathway towards the evolution of reduced dispersal capacity – increases in seed size rather than reductions in the functionality of dispersal aides – will be referred to as the *size-constraints hypothesis*.

Carlquist (1966a, 1966b) clearly recognised this mechanistic pathway towards reduced dispersibility: "Extreme cases of loss of dispersability ... feature increase in fruit size without concomitant increase in appendages which serve in dissemination" (Carlquist 1966b, p. 46). Furthermore, he reasoned, "a poorly functioning dispersal apparatus may be retained because it does not have a strongly negative selective value" (Carlquist 1966a, p. 47).

He also noticed something peculiar about many island endemics with apparently smaller dispersal aides, particularly species in the genus *Bidens*: Asteraceae (Fig. 3.2). In North America, seeds produced by *Bidens* spp. typically have hooked awns, which stick tenaciously to the feathers and fur of animals. The genus also occurs throughout the Pacific and was presumably dispersed to islands in Oceania stuck to the feathers of seabirds. At first glance, the morphology of *Bidens* spp. across the Pacific appears to support the loss of dispersibility hypothesis, as many island endemics have much smaller awns than related species on the mainland (Fig. 3.2). Yet, strangely, many species endemic to the Hawai'ian Islands "have large flat achenes which can sometimes be said to be winged" (Carlquist 1966b, p. 35). Other island endemics, including

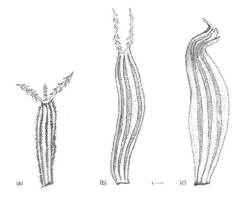

FIGURE 3.2 Variation in fruit morphology in the genus *Bidens*: Asteraceae. (a) *Bidens pilosa*, North America; (b) *Bidens skottsbergii*, Hawai'i; (c) *Bidens magnidisca*, Hawai'i). Island species tend to have reduced awns and larger, flattened seeds. Redrawn from Carlquist (1974). The scale is approximately 1 mm.

species in the genus *Oparanthus*: Asteraceae, which occurs in the Marquesas and Rapa Nui, as well as *Lecocarpus pinnatifidus*: Asteraceae, which inhabits the Galapagos, also possess secondarily derived wings on the flanks of their seeds. *In situ* evolution of winged seeds seems strange from a shipwrecked mariner's perspective. If selection favours the loss of dispersibility in plants, why have many island endemics evolved larger, apparently novel, dispersal aides?

In an insightful investigation into the aerodynamics of autorotating samaras, Greene and Johnson (1993) showed that reductions in dispersal capacity can evolve via selection for larger overall fruit size. They demonstrate that if selection favours larger seeds, reduced dispersal capacity arises passively when the relationship between seed size and samara size remains constant (i.e., isometric). Therefore, reductions in dispersal potential can result from selection for large fruits, for whatever reason, even when the size of wings also evolve to become bigger (Fig. 3.3).

Previous work on the loss of dispersibility rarely distinguishes between the mechanisms that might generate reductions in dispersal potential. When a loss of dispersibility is documented in an island plant population, does it arise from increases in seed size, reductions in dispersal aides, or changes in both? Answering this question is important.

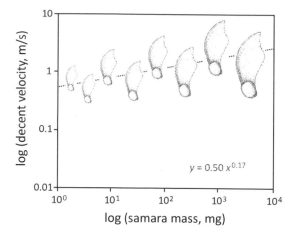

FIGURE 3.3 Relationship between fruit mass and decent velocity in auto-rotating samaras (redrawn from Green & Johnson 1993). Decent velocity, which is inversely related to dispersal potential, increases with fruit mass, even when the relationship between seed size and wing size remains constant.

Although the loss of dispersal potential has traditionally been argued to arise directly from selection acting on dispersal aides, as Darwin alluded to, it could just as easily arise as a passive by-product of selection for large seeds, for reasons that may be wholly unrelated to their dispersal.

This chapter reviews the literature and presents new data to investigate whether the loss of dispersal potential is a repeated pattern in the evolution of island plants. It focuses on available data that allow definitive tests of both the loss of dispersal and size constraints hypotheses, by reporting information on the proximate basis of dispersal – seed size and the functionality of dispersal aides. Afterwards, it reviews other studies that provide relevant information, but lack sufficient data to test each hypothesis directly.

HYPOTHESIS TESTING

Complete Tests of Predictions

Mycelis muralis *and* Hypochaeris radicata *in the Northeast Pacific*
Wall lettuce (*M. muralis*: Asteraceae) and cat's ear (*H. radicata*: Asteraceae) are weedy herbs native to Europe. However, they have been

(a)

(b)

FIGURE 3.4 (a) *Mycelis muralis*: Asteraceae fruits and (b) islets on the west coast of Vancouver Island, British Columbia, Canada, the location of Cody and McC. Overton's (1996) influential study. (photo from Getty Images)

introduced across the globe by human activities and they now occur in some of the most remote locations on the planet outside of their natural range. Like many species in the Asteraceae, both species produce wind-dispersed fruits consisting of a seed attached to a fluffy plume (Fig. 3.4).

In a long-term study spaning decades, Cody and McC. Overton (1996) documented the demography and fruit morphology of *M. muralis* and *H. radicata* on islets in a remote glacial fjord off the west coast of Canada (see also Cody 2006). They repeatedly censused island populations over time, documenting when island populations became established, when they went extinct, and how populations changed in size over time. They also quantified the morphology of fruits produced by island populations of different ages to test the loss of dispersibility hypothesis. More specifically, they measured plume volumes, seed volumes, and the time it took fruits to reach the ground when released from above (i.e., their decent velocities).

Results showed that island populations of *M. muralis* turned over rapidly, with colonisation and extinction events occurring frequently. When islands were first colonised, *M. muralis* populations were typically comprised of one or a small number of individuals. Afterwards, most island populations increased steadily in size, resulting in a positive relationship between the age and size of island populations. A thorough test of the loss of dispersibility hypothesis could therefore be made with *M. muralis*, namely that younger populations would display relatively good dispersibility, and that older populations would display a progressive loss of dispersal capacity. On the other hand, *H. radicata* populations turned over more slowly on islands. The size and age of their island populations were unrelated, so comparisons of fruit morphology could only be made between mainland populations and older, well-established island populations. Cody and McC. Overton (1996) also found that the ratio between the volume of plumes '*P*' and the volume of seeds '*S*', which they referred to as the '*DISPER*' ratio (*DISPER* = *P*/*S*), was inversely related to the descent velocity of both species, and was therefore an accurate reflection of each fruit's dispersal potential.

Consistent with the loss of dispersibility hypothesis, DISPER ratios in *M. muralis* populations varied with their age. Younger populations had higher values of DISPER, indicating they generally had greater powers of dispersal. On the other hand, older island populations had lower DISPER ratios, suggesting island selection pressures favoured the evolution of reduced dispersal potential in more established populations. Similarly, comparisons of DISPER ratios in *H. radicata* showed that well-established island populations generally had reduced dispersal potential relative to mainland populations.

Closer inspection of the two variables that jointly determine DISPER ratios hint at the mechanisms responsible for reductions in dispersal capacity. Plume volumes showed weak, statistically insignificant declines with population age in both species. On the other hand, seed size increased markedly on islands in both species. Reductions in dispersal potential (DISPER) therefore resulted mostly from increases in seed size. Because selection may favour increased seed size for reasons that are unrelated to dispersal potential, declines in dispersal potential in *H. radicata* and *M. muralis* may not necessarily result from selection for reductions in dispersal potential *per se*. Given that population densities tended to increase with population age, perhaps reductions in dispersal potential resulted from density-dependent selection for large seeds, which give rise to larger, more competitive seedlings (Geritz et al. 1999; Coomes & Grubb 2003; Lönnberg & Eriksson 2013).

Similar patterns in dispersal potential of *M. muralis* have been observed in fragmented habits in continental Europe. Riba et al. (2009) measured plume and seed morphology across Southern Europe and found that DISPER ratios increased with landscape connectivity and declined with habitat fragmentation, which is generally consistent with Cody and McC. Overton's (1996) findings on Canadian islands. However, values of plume and seed sizes were not reported independently in this study, so it remains unclear whether reductions in dispersal potential arose from changes in seed size, plume size, or both.

Cirsium arvense and Epilobium spp. in the Danish Straits
Creeping thistle (*C. arvense*: Asteraceae) is a common European weed that produces wind-dispersed fruits comprised of a small seed attached to a feathered plume, similar to those produced by *H. radicata* and *M. muralis*. Willowherbs (*Epilobium angustifolium* and *Epilobium hirsutum*: Onagraceae) produce functionally analogous fruits, but in these species their plumes are derived from different anatomical structures. All three perennial herb species inhabit small islands in the Danish Straits, a relatively small body of water separating the North Sea from the Baltic Sea in Northern Europe.

Fresnillo and Ehlers (2008) investigated the fruit morphology of *C. arvense*, *E. angustifolium* and *E. hirsutum* on three islands in the Danish Straits: Tuno (6.0 km off the Danish coast), Endelave (9.5 km offshore), and Anholt (45.0 km offshore). They measured DISPER ratios, descent velocities and fruit weights of island populations and compared them to mainland populations to test the loss of dispersibility hypothesis.

Results showed that island populations of *C. arvense* had faster descent velocities relative to mainland populations, indicating that island plants tended to have lower dispersal potential than mainland plants (Fresnillo & Ehlers 2008). Fruit weights were also higher in island plants, suggesting that faster descent velocities could result from increases in seed size, rather than declines in plume size. In *E. angustifolium*, island populations tended to have slower descent velocities, even though they had heavier fruit weights. Therefore, rather than having reduced dispersal potential, island plants actually had greater dispersal potential than mainland plants. Similar to *E. angustifolium*, descent velocities in *E. hirsutum* tended to be slower in island populations, indicating that island plants had greater dispersal potential than mainland plants. However, in this species seed weights were similar between island and the mainland populations.

Overall, Fresnillo and Ehlers' (2008) findings differed strongly among their three study species. One species showed evidence for reduced dispersal potential on islands. However, reductions in dispersal potential in this species were associated with increased seed mass, indicating that processes other than selection for reduced dispersal potential could be responsible. The other two study species actually had higher dispersal potential on islands, one of which also had heavier seeds on islands.

Periploca laevigata *in Macronesia*

Wolfbane (*P. laevigata*: Apocynaceae) is an anemochorous shrub that occurs in Southern Europe, islands in the Atlantic (e.g., Canary and Cape Verde Islands) and islands in the Mediterranean. It produces seed pods that dehisce to release multiple seeds attached to feathery plumes.

García-Verdugo et al. (2017) conducted a particularly thorough study of the dispersal potential of *P. laevigata*. They measured the size of seeds and their associated plumes, as well as their descent velocities, on islands in the Atlantic and the Mediterranean, as well as mainland

Europe. They conducted molecular analyses to establish the age of island populations in order to test whether dispersal potential declines with population age, as predicted by the loss of dispersal hypothesis. They also conducted extensive glasshouse experiments to determine how fruit morphology might be influenced by environmental conditions.

Results showed no evidence for a loss of dispersal potential on islands. Instead, plants inhabiting isolated islands in the Atlantic Ocean actually had higher dispersal potential than plants in continental Europe, as well as islands in the Mediterranean. The dispersal capacity of *P. laevigata* fruits was unrelated to the age of island populations and trends in dispersal capacity persisted in common garden experiments. Plants on both the Canary and Cape Verde Islands produced bigger seeds than plants on the mainland and on islands in the Mediterranean, a trend that also persisted in standardised conditions.

Crepis sancta *in Montpellier, France*

Hawksbeard (*C. sancta*: Asteraceae) is a weedy herb native to Europe and Western Asia that displays a marked dimorphism in the morphology of its fruits. Like most species in the Asteraceae, it produces composite flowers comprised of disk and ray florets. Seeds arising from disk florets are attached to fluffy plumes that promote wind dispersal. Seeds arising from ray florets lack a plume and typically fall within the immediate vicinity of parent plants (i.e., barochory).

Cheptou et al. (2008) investigated spatial variation in the production of dispersive seeds (i.e., plumed seeds arising from disk florets) versus sedentary seeds (i.e., plumeless seeds arising from ray florets). Working in urban areas of Montpellier, France, they documented the occurrence of *C. sancta* in small patches of soil beneath trees planted along city sidewalks. Soil beneath urban trees provides island-like patches of inhabitable space for weeds, separated from each other by a matrix of uninhabitable sidewalk.

They measured the relative frequencies of dispersive and sedentary seeds across multiple habitat patches to test the hypothesis that sedentary seeds would be produced in higher frequencies beneath isolated urban trees. They also established the relative costs of dispersal to uninhabitable cement sidewalk, which is analogous to 'sea-swept' costs of dispersal of plants inhabiting true islands.

Results showed that dispersive seeds were more likely to settle in unsuitable concrete habitat than sedentary seeds. In fact, dispersive seeds were 55% less likely to settle in their patch of origin than non-dispersive seeds (Cheptou et al. 2008), indicating the costs of dispersal were lower in sedentary seeds. Concomitant with the relative costs of dispersal, urban plants tended to produce greater proportions of sedentary seeds, suggesting that selection has favoured this fruit form in as little as 5–12 generations (Cheptou et al. 2008).

In an earlier study of *C. sancta*, Imbert (1999) found that dispersive and sedentary seeds differed strongly in size. Sedentary seeds were more than twice as heavy as dispersive seeds (0.27 mg vs 0.10 mg), and their embryos were also more than twice as large (0.15 mg vs 0.06 mg). Furthermore, Imbert et al. (1996) showed that differences in seed and embryo size between morphs were consistent among populations at different stages of succession.

Rumex bucephalophorus *in the Atlantic and Mediterranean*

Red dock (*R. bucephalophorus*, Polygonaceae) is an annual herb that inhabits coastal regions in the Mediterranean basin. Its vegetative growth form is highly variable – some plants produce leaves in basal rosettes, while others produce leaves on aerial stems. Its fruit and seed morphology are also highly polymorphic.

Four categories of fruit and seed morphology are apparent in *R. bucephalophorus* (see Talavera et al. 2012). The first category is a typical wind-dispersed fruit, consisting of a seed attached to a wing that aides in wind dispersal. The second is gravity-dispersed fruits, which are similar to the first, but lacking a wing. A third type of fruit remains attached to parent plants after they die at the end of the growing season and are thereby rooted to their site of origin and are largely sedentary. The last type of dispersal structure is produced only by plants that form rosettes. This type of fruit originates at the junctions of leaves and are eventually buried by contractile roots (i.e., geocarpy). Consequently, *R. bucephalophorus* exhibits two general modes of fruit form, dispersive seeds that detach from parent plants (i.e., winged and unwinged seeds) and sedentary seeds that are connected to the substrate in the immediate vicinity of their parent plants.

Talavera et al. (2012) compared of the relative frequencies of dispersive and sedentary fruit types among populations in continental Europe, islands in Macaronesia (the Canary Islands and Madeira) and islands in the Mediterranean (Corsica, Sardinia, Sicily, and Balearic Islands). Results showed that although the relative frequencies of different dispersal modes in *R. bucephalophorus* varied widely, consistent differences in dispersal potential between islands and the mainland were not observed.

Sonchus *in the South-west Pacific*
Chatham Island sow thistle (*Sonchus grandifolia*: Asteraceae) is a perennial herb that is endemic to the Chatham Islands (Fig. 3.5). Like many other species inhabiting high-latitude islands in the South Pacific, *S. grandifolia* is a sub-Antarctic 'megaherb'. It is far larger than its closest ancestors, both of which occur in New Zealand (*Sonchus kirkii* and *Sonchus novae-zelandiae*, Wagstaff & Breitwieser 2002). Similar to most other species in the Asteraceae, they produce wind-dispersed fruits that are comprised of seeds attached to fluffy plumes. Comparisons of fruit morphology showed that the island endemic has reduced dispersal potential (Box 3.1). However, differences in dispersibility result from increases in seed size rather than declines in plume volume.

FIGURE 3.5 *Sonchus grandifolia*: Asteraceae, a sub-Antarctic 'megaherb' from the Chatham Islands, which exhibits reduced dispersal potential resulting from seed gigantism, rather than reductions in the size of dispersal aides (feathery plumes).

BOX 3.1   **Seed gigantism and loss of dispersibility in**
Sonchus

Sonchus grandifolia: Asteraceae is a sub-Antarctic mega-herb that is
endemic to the Chatham Islands, which shares a recent common
ancestor with *Sonchus kirkii*, a related species that is endemic to New

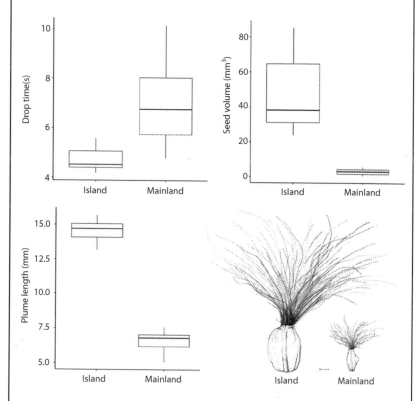

FIGURE B3.1 Fruit trait comparisons between *Sonchus grandifolia*:
Asteraceae, which is endemic to the Chatham Islands, and *Sonchus
kirkii* a close relative on the New Zealand 'mainland'. The island
species had faster drop times, bigger seeds, and longer plumes than
the mainland species. However, increases in seed size were more
pronounced than increases in plume size, indicating that reductions
in dispersal potential result from disproportionate increases in seed
size. The scale (lower right) represents 1 mm.

BOX 3.1   **(cont.)**

Zealand. To test the loss of dispersibility and size constraints hypotheses, descent velocity, seed volume, and plume length were measured and compared between *S. grandifolia* and *S. kirkii.*

Twenty fruits of each species were collected from the field, released from 2 m above the ground in still conditions, and the time it took for each fruit to descend to the ground was noted. Each fruit was dropped three times and results were averaged to obtain a single value for each fruit. Seed volumes and plume lengths were measured using a dissecting scope. All three variables were compared between species with linear models in the *R* environment (R Core Team 2014). Decent times and seed volumes were log transformed to conform to assumptions.

Results showed that of *S. grandifolia* had faster drop times than *S. kirkii* (t = 7.416, $p < 0.001$), indicating that the island endemic exhibited a loss of dispersal capacity (Fig. B3.1). However, contrary to the loss of dispersibility hypothesis, *S. grandifolia* actually had longer plumes than *S. kirkii* (t = 37.880, $p < 0.001$). Instead, *S. grandifolia* produced bigger seeds than *S. kirkii* (t = 15.040, $p < 0.001$), consistent with the size constraints hypothesis. Therefore, reductions in dispersal capacity in *S. grandifolia* seeds were generated by disproportionate increases in seed size relative to plume length. The data were courtesy of Cherie Balls.

Hibiscus tiliaceus *in the South-west Pacific*
Beach hibiscus (*H. tiliaceus*: Malvaceae) is a tree that grows in coastal areas of the Pacific and Indian Oceans. Unlike the previous examples, its seeds contain air spaces that enable them to float in salt water and are therefore water dispersed (Kudoh et al. 2006). Extensive genetic analyses of populations distributed across its geographic range have revealed little genetic differentiation (Takayama et al. 2006). So long-distance dispersal across large stretches of open ocean appears to occur quite frequently (see also Kudoh & Whigham 2001). Therefore,

H. tiliaceus appears to be a single panmictic species across most of the world's oceans (Takayama et al. 2006).

A notable exception is *Hibiscus glaber*, a closely related taxon that occurs on the Ogasawara Islands. Kudoh et al. (2006) quantified differences in the distribution of *H. tiliaceus* and *H. glaber* between inland and coastal habitats, in addition to conducting seed dissections and buoyancy trials (see also Kudoh et al. 2013). Although clearly derived from *H. tiliaceus* (Takayama et al. 2005), which typically occurs along the coast, *H. glaber* is more commonly found away from the coast in forested habitats. Seed dissections showed that *H. glaber* seeds typically lack the air spaces produced by *H. tiliaceus*. As a result, *H. glaber* seeds usually sink in saltwater, while *H. tiliaceus* seeds float (Kudoh et al. 2006). Therefore, the evolutionary loss of air spaces in *H. glaber* seeds may help promote its more inland distribution by reducing the likelihood of being washed out to sea by rainwater.

Seed dissections also revealed that the air spaces in *H. tiliaceus* seeds are formed by a reduction in the size of their cotyledons. Conversely, the loss of air pockets in *H. glaber* results from greater investment cotyledons. Therefore, reduced buoyancy in *H. glaber* seeds appears to be linked to greater seedling provisioning.

Partial Tests of Predictions

The aforementioned examples provide not only estimates of dispersibility in island plants, but also estimates of seed or embryo size, as well as measurements of the size of dispersal aides. They can therefore be used to test both the *loss of dispersibility* and *size-constraints hypotheses*. The following examples provide useful information on fruit form in island plants, but lack enough detail for tests of both hypotheses.

Sophora *in the Pacific*

The genus *Sophora* sect Edwardsia: Fabaceae encompasses approximately 19 species of trees and shrubs that occur mostly on islands in the Pacific. Eight species occur in New Zealand (Fig. 3.6) and multiple

FIGURE 3.6 Flowers and seed pods of *Sophora microphylla*: Fabaceae from New Zealand.

taxa occur in both Chile and the Juan Fernandez Islands. *Sophora* species also occur in Hawai'i, Easter Island, Lord Howe Island, French Polynesia, Reunion Island in the Indian Ocean, and Gough Island in the Atlantic Ocean.

Most species in the genus *Sophora* sect Edwardsia show limited genetic subdivision (Hurr et al. 1999; Shepherd & Heenan 2017a, 2017b; Shepherd et al. 2017). Only a few are genetically isolated, most notably *Sophora chrysophylla* in Hawai'i and *Sophora howinsula* on Lord Howe Island. Instead, many species, such as *Sophora microphylla* in New Zealand and *Sophora cassioides* in Chile, show evidence of genetic mixing over large geographic distances (Shepherd & Heenan 2017b).

All *Sophora* species produce large seeds within fibrous seed pods. Although the seed pods float, when exposed to seawater they decompose within weeks and release the seeds. The seeds of some *Sophora* species can float for years and readily germinate afterwards, so extensive water dispersal may explain low levels of genetic subdivision across its geographic range (Sykes & Godley 1968). However, the mechanisms underpinning seed buoyancy have yet to be firmly established. Several attributes may promote seed buoyancy, including air pockets, the density of embryonic tissues (Guppy 1906), or by waxy seed coats (Thorsen et al. 2009). A better understanding of the proximate mechanisms underpinning seed buoyancy precludes a rigorous test of the loss of dispersibility and size-constraints hypotheses.

Sophora seeds regularly wash ashore on islands where adult plants do not occur. For example, seeds have been collected from beaches on the Kermadec Islands, approximately 1,000 km from the nearest population of adult trees in New Zealand. Molecular analyses of these beached seeds indicate that they could have come from New Zealand. However, longer-distance dispersal from other populations in the Pacific could not be ruled-out (Shepherd & Heenan 2017a).

In 2010, a seed from an unknown species of *Sophora* was found at the edge of an interior lake on Macquarie Island (Smith 2012). Given its close proximity to Antarctica, Macquarie Island only supports herbaceous plants so, much like the seeds collected from the Kermadec Islands, this seed must have arrived on the island via long-distance dispersal. However, it's occurrence in a landlocked lake suggests dispersal vectors other than water dispersal might be at work. Although the seed could have blown inland after a long journey floating on the open ocean, it could also have been dispersed inland after being consumed by a bird. Many species of seabird ingest objects floating on the sea surface that presumably resemble food (Ryan 1987). Given global increases in plastic pollutants on the surface of the sea, this behaviour currently threatens the well-being of many seabirds (Wilcox et al. 2015). However, it might also aide in the dispersal of floating seeds.

Interestingly, not all species of *Sophora* produce buoyant seeds. In a pioneering study of hydrochory, Guppy (1906) conducted floatation trials on *Sophora* seeds collected in Chile (*S. cassioides*) and Hawai'i (*S. chrysophylla*). Although he did not report precise sample sizes, he found that all of the seeds he collected along the Chilean coast stayed buoyant for considerable lengths of time. Conversely, all of the seeds he collected from the Hawai'ian species, which is endemic to high-elevation forests, sank when immersed in water. Similar seed trials on *S. howensia*, which inhabits rainforests of Lord Howe Island, revealed no evidence for seed buoyancy (Ian Hutton personal communication, N = 200). Intermediate levels of seed buoyancy were observed in *S. microphylla* (22%, N = 690; 11.7%, N = 1,200; 52.5%, N = 232), a widely distributed taxon across a range of habitats of in New Zealand.

These buoyancy trials suggest that the capacity for water dispersal declines in taxa that are restricted to inland habitats on oceanic islands. However, there is no evidence for systematic variation in seed size. Populations in Chile (seed length = 7–8 mm, Rix & Lewis 2016), Hawai'i (seed length = 4–8 mm, Wagner et al. 1999), Lord Howe Island (seed length = 7mm, Green 1970), and New Zealand (seed length = 5–8 mm, Allan 1982) produce similarly sized seeds.

Island Bryophytes

Bryophytes commonly occur on isolated islands. However, unlike seed plants, they disperse to new locales as spores. Spores are functionally similar to seeds except that they lack obvious dispersal aides, such as wings, awns, or air bladders. So their dispersal potential is determined directly by their size, with smaller, lighter spores having greater dispersal potential than larger, heavier spores. Bryophytes produce two types of spores – large asexual diaspores with shorter dispersal distances, and small sexual diaspores with greater dispersal distances.

Island bryophytes reproduce more frequently with large, asexual diaspores than their mainland counterparts (Patiño et al. 2013). This indicates that island bryophytes generally have reduced dispersal potential. However, a variety of other factors could select for differences in spore size aside from their dispersibility. For example, large spores could have greater establishment probabilities. Alternatively, because large diaspores result from asexual reproduction and small diaspores result from sexual reproduction, selection for different mating systems on islands could also result in the loss of dispersal potential (i.e., Baker's law, Chapter 4).

Canary Island Endemics

Vazačová and Münzbergová (2014) compared the dispersal potential of a large number of plant species endemic to the Canary Islands to co-occurring, non-endemic congeners. Results failed to show consistent support for the loss of dispersal potential, regardless of dispersal mode. However, unlike previous examples, comparisons were made between endemic species and congeners that also occur on the Canary Islands. Therefore, differences in fruit morphology between species pairs could result from 'immigrant selection', rather than natural selection for increased seed size or reduced functionality of dispersal aides, given that non-endemic species likely arrived on the Canary Islands more recently, from ancestors that may have had relatively high dispersal potential.

CONCLUSIONS

The loss of dispersibility in island populations has intrigued evolutionary biologists for well over a century. However, quantitative tests for the loss of dispersibility in island plants are rare. Only ten island–mainland comparisons could be found that assessed whether differences in dispersal resulted from changes the size of seeds or dispersal aides (Table 3.1). Reductions in dispersibility were observed in six comparisons, while the opposite trend, higher dispersal potential in insular environments, was observed in three comparisons. Consequently, the loss of dispersal potential does not appear to be a highly repeated pattern in the evolution of island plants.

Despite the appeal of the loss of dispersibility hypothesis, empirical support is equivocal. When reduced dispersal potential is documented, it often involves increased seed size. This indicates that the loss of dispersal potential could evolve as a passive by-product of selection for large seeds, for reasons that are wholly unrelated to their

Table 3.1 *Summary of evidence for two explanatory hypotheses for the reduction of dispersal potential in island populations*

	Reduction in dispersibility?	Decline in **dispersal aides?**	Increase in **seed/ embryo size?**
Hypochaeris radicata[a]	Yes	No[1]	Yes
Mycelis muralis[a]	Yes	No[1]	Yes
Cirsium arvense[b]	Yes	Yes	Yes
Epilobium angustifolium[b]	No[3]	No	Yes
Epilobium hirsutum[b]	No[3]	Yes	No
Periploca laevigata[c]	No[3]	No[4]	Yes[5]
Crepis sancta[d]	Yes	Yes[2]	Yes
Rumex bucephalophorus[e]	No	No	No[6]
Sonchus spp.[f]	Yes	No[4]	Yes
Hibiscus spp.[g]	Yes	Yes	Yes

A reduction in dispersal potential coupled with a reduction in dispersal structures would be consistent with the loss of dispersibility hypothesis. Changes in dispersal potential coupled with changes in seed (or embryo) size would be consistent with the size-constraints hypothesis. [1] Weak (statistically insignificant) decline in plume size, [2] increased frequency of fruits without plumes, [3] dispersal potential higher in insular environments, [4] plume sizes higher in insular environments, [5] seed mass higher on islands in the Atlantic but not islands in the Mediterranean, [6] seed sizes were smaller on islands, [a] Cody & McC. Overton (1996), [b] Fresnillo & Ehlers (2008), [c] García-Verdugo et al. (2017), [d] .Cheptou et al. (2008), [e] Talavera et al. (2012), [f] Ball & Burns (unpublished), [g] Kudoh et al. (2013)

dispersal. As a result, a fresh approach to the study of how evolution shapes the evolution of dispersal potential on islands might be needed in the future (Burns 2018).

Perhaps more than ever, the subject is an interesting and important topic for ongoing research. Carlquist (1966a, 1966b, 1980) discusses a large number of putative examples of seed gigantism and the loss of dispersal potential in island plants, all of which are worthy of quantitative study. Given that the majority of quantitative examples to date have been conducted on anemochorous species, future study would benefit from focusing on other dispersal mechanisms (e.g., hydrochory and epizoochory). Perhaps most importantly, testing for differences in the dispersal potential of island plants and their mainland relatives is just the first step in understanding how insularity shapes the evolution of fruit form. Delving deeper into the mechanistic basis of differences in dispersal potential is critically important, as it will help resolve whether selection is acting directly on dispersal potential, or whether dispersal potential evolves as a passive by-product of selection acting on other plant parts.

4   Reproductive Biology

It is not often that research on an evolutionary topic carried out independently by botanists and zoologists produces conclusions which are virtually identical. When this does happen one cannot restrain a feeling that a principle of more than superficial importance has been uncovered.

H. G. Baker (1955)

In 1955 Herbert Baker wrote a brief commentary in the journal *Evolution* highlighting the similarities between his work on leadworts (Plumbaginaceae), and the work of A. R. Longhurst, a zoologist working on the reproductive biology of tadpole shrimp (Notostraca). Both men noticed that self-compatibility in their respective study organisms was more prevalent in recently established populations. They therefore came to the same conclusion – self-compatibility is somehow linked to dispersal.

Long-distance dispersal to an isolated patch of suitable habitat, like an oceanic island, is an improbable event. So when it does occur, it is likely to be accomplished by just one or only a small number of individuals. Self-compatibility would be an obvious advantage under these conditions. Freed from the need of another individual to reproduce, self-compatible organisms can initiate a new population all on their own.

Several years later, Stebbins (1957) referred to this association as 'Baker's law'. To be clear, he was not trying to say that the relationship between dispersal and self-compatibility was unbreakable, like a basic law in physics (e.g., Newton's law of universal gravitation). Stebbins meant that it was a repeated pattern or an ecogeographic 'rule' (see Colyvan & Ginzburg 2003; Gaston et al. 2008). Regardless of terminology, Baker's idea has guided research into spatial patterns in sexual reproduction ever since. It has also proved contentious.

A decade after Baker's observation that self-compatibility and long-distance dispersal were linked, an apparent exception was noticed by Sherwin Carlquist (1966a; 1966b; 1966c). He highlighted that in some oceanic archipelagos, and in particular Hawai'i, dioecy appeared to be unusually common. Dioecious species are by definition self-incompatible. So, if gender dimorphism, in general, and dioecy, in particular, was shown to be over-represented on islands, it would directly contradict Baker's law.

Baker (1967) responded to Carlquist's observation by suggesting that the two phenomena, self-compatibility and gender dimorphism, may both be favoured on isolated islands, albeit at different points in time. He defended his law by arguing that self-compatibility should only be favoured at the initial stages of population establishment. After an island population was fully established, the tables would turn and outcrossing mechanisms such as dioecy would be favoured. In other words, once a founding population of self-compatible individuals was large enough to eliminate mate limitation, self-incompatibility would be favoured to reduce inbreeding depression (see Pannell 2015).

This series of demographic and genetic changes following island colonisation can be formalised into a single, time-staged prediction for the reproductive biology of island plants. At the initial stage of island colonisation, immigrant selection favours self-compatibility. Later, as mate limitation declines following population establishment, natural selection favours the evolution of traits that promote outcrossing, such as dioecy and heterostyly, to avoid inbreeding depression (Baker 1967; Thomson & Barrett 1981; Bawa 1982; Pannell et al. 2015).

Another long-standing paradigm in the reproductive biology of island plants involves the size and shape of flowers produced by island plants. Naturalists have long recognised that island flowers are remarkably unremarkable (Hooker 1853; Wallace 1876; Carlquist 1974; Lloyd 1985). In contrast to continental plant communities, which are typically filled with many species that produce large, colourful flowers, island floras are dominated by small, unpigmented flowers. Despite the long history of speculation surrounding the morphology of island flowers, quantitative tests for overarching trends in the size of island flowers are rare.

What factors might consistently shape island flowers? One possibility is that insular changes in floral morphology are linked to pollination mode. While the diversity of animal pollinators may be lower on islands, wind is not (Box 1.2). Selection might therefore favour anemophily, or wind-pollinated flowers, over animal-pollinated flowers. Freed from selection for energetic rewards and conspicuous displays to attract animal pollinators (Johnson et al. 1995), wind-pollinated flowers are often small, unpigmented, and inconspicuous. Alternatively, many types of pollinators that are common on continents are rare or absent on islands (see Barrett 1996; Traveset & Navaro 2018), including hawkmoths (Sphingidae), butterflies (Papilionoidea), and long-tongued bees (Apidae). In their absence, generalist pollinators such as flies (Diptera) commonly visit flowers (Lloyd 1985) and generalist pollinators may select for flowers that lack structures suited specifically to specialised pollinators. Reduced diversity in the size of pollinator faunas might also result in reduced diversity in the size of island flowers.

SYNDROME PREDICTIONS

This chapter investigates several aspects of the reproductive biology of island plants. First, it summarises the evidence to date for Baker's law, which states that the capacity for self-compatibility is over-represented on isolated islands. Next, evidence for the opposite trend is explored, whether dioecy might instead be more prevalent on islands. This prediction was tested with a brief summary of the literature on gender dimorphism on islands and a new, global-scale analysis of insular dioecy. A review of the evidence for a preponderance of heterostyly on islands is also conducted. Three aspects of the floral biology of island plants are explored. Assuming that pollinator communities on islands are less diverse than pollinator communities on continents, three predictions can be made of flowers on islands. (1) Anemophily is more common on islands. (2) Island flowers tend to lack specialised structures associated with particular pollinators and are therefore more generalised. (3) The diversity of flower sizes displayed by island plant communities is lower than the diversity of flower sizes in continental plant communities. To test these three predictions, past research on anemophily and generalised flowers is reviewed, and a novel analysis of flower size diversity on islands in the South-west Pacific is presented to test for evidence of reduced flower size diversity on islands.

HYPOTHESIS TESTING

Baker's Law

Tests for increased self-compatibility on islands have not always provided clear-cut support for Baker's law (de Waal et al. 2014; Pannell 2015; Pannell et al. 2015). A qualitative review of self-compatibility by Newstrom and Robertson (2005) failed to show consistent differences between the New Zealand flora and elsewhere. A more recent quantitative comparison of self-compatibility on sub-Antarctic Islands also failed to provide definitive support for Baker's law (Lord 2015). However, several other recent studies have clarified the situation by conducting global, rather than regional, tests of Baker's law. Igic et al. (2008) found that self-compatibility is more prevalent on Chiloé Island, the Galapagos Islands, Juan Fernández Island, Jamaica, and New Zealand than it is on continents. In an impressively comprehensive analysis of the distribution of hundreds of island bryophytes distributed across the globe, Patiño et al. (2013) found that the proportion of species reproducing bisexually was higher on islands relative to continents.

In the most comprehensive meta-analysis to date, Grossenbacher et al. (2017) reviewed all previous studies testing Baker's law over the past 50 years. They considered hundreds of studies on thousands of plant species on continents and on islands in the Atlantic, Indian, and Pacific Oceans. From these studies, Grossenbacher et al. (2017) assembled a standardised data set from previous work focusing specifically on

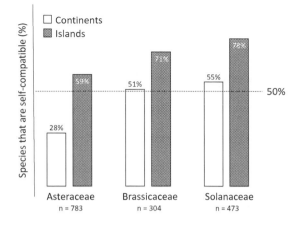

FIGURE 4.1 Differences in the frequency of self-compatibility between islands (grey bars) and continents (white bars). Analyses are restricted to three diverse plant families: Asteraceae, Brassicaceae, and Solanaceae. Exact percentages of species that are self-compatible are shown above and total sample sizes of species from each family are shown below. Data from Grossenbacher et al. (2017)

self-compatibility. They restricted their attention to three specious families (Asteraceae, Solanaceae, and Brassicaceae) to determine whether trends were phylogenetically independent. Similar trends were obtained from all three families. As predicted by Baker's law, self-compatibility was over-represented on islands relative to continents (Fig. 4.1).

Other previous work has explored the proximate mechanisms underpinning greater incidence of outcrossing mechanisms in colonising populations. For example, Hesse and Pannell (2011) found that in spatially isolated conditions, strictly female plants set fewer seeds than hermaphrodites. However, this density-dependent effect on outcrossing disappeared in more crowded conditions, where hermaphrodites and females set similar amounts of seed (see also Ison & Wagenius 2014; Pannell 2015).

Dioecy

Gender dimorphism, and in particular dioecy (Fig. 4.2), has often been hypothesised to be more prevalent on isolated islands (see Grant 1998;

(a)

FIGURE 4.2 Separate (a) male and (b) female flowers of *Coprosma huttonia*: Rubiaceae, a dioecous plant species endemic to Lord Howe Island. Note the pollen collecting on the leaves below the male flowers. Photo taken by Ian Hutton

(b)

Whittaker & Fernández-Palacios 2007). In an attempt to clarify conflicting reports that both self-compatibility and dioecy are over-represented on islands, Baker and Cox (1984) compiled available information on the incidence of dioecy in insular plant communities (n = 22) and in continental plant communities (n = 5). Results showed that the incidence of dioecy was similar between islands and continents. Interestingly, further analyses of the 22 insular sites showed strong geographic patterning in dioecy. Over 80% of among-island variation in the incidence of dioecy could be attributed to the geography of islands. Dioecy increased with island elevation and declined with latitude, indicating that taller, tropical islands housed greater frequencies of dioecious species than low-lying, temperate islands. They conclude that dioecy varies geographically with environmental conditions, rather than insularity *per se*.

Building upon Baker and Cox's (1984) earlier analyses, Box 4.1 presents the results of a more extensive, global-scale analysis of the distribution of dioecy. A literature search was conducted to locate studies reporting on the incidence of dioecy in plant assemblages from around the globe and the resulting data were used to test whether dioecy varies with precipitation, temperature, latitude, or insularity. Results showed that the global incidence of dioecy increased with precipitation, but was unrelated to temperature, latitude, and insularity. Wetter areas contain more dioecious species, both on islands and on continents. In fact, there was a weak statistical tendency for dioecy to be less common on islands. Therefore, previous speculation that dioecy is more prevalent on islands may have arisen from a confounding relationship with precipitation. Dioecy, therefore, appears to be common on islands like Hawai'i and New Zealand because they receive high levels of precipitation, not because they are islands.

Why would dioecy increase with precipitation? Dioecy has often been linked to other plant traits (see Sakai & Weller 1999). Therefore, increases in the incidence of dioecy with precipitation might result from associations with other traits that are favoured in wetter environments. For example, one explanation for the results in Box 4.1 involves a cascade of relationships among several life history correlates of dioecy, including outcrossing, seed size, and dispersal mode. Dioecy has an obvious effect on outcrossing, as strictly dioecious species are incapable of self-pollination. Outcrossing, in turn, influences seed size. A literature search for studies investigating how outcrossing affects seed sizes (Supplemental Resources 4.2) showed that 20 out of 26 studies (78%) found a positive effect of outcrossing on seed sizes. As seed size increases, it imposes a severe constraint on the aerodynamics of wind dispersal so large seeds are typically fleshy-fruited and dispersed by animals (Westoby et al. 1996; Leishman et al. 2000; Herrera 2002, see also Muenchow 1987). Finally, fleshy fruits are distributed non-randomly across the globe, being over-represented in wet

---

BOX 4.1   **Insular dioecy**

To test whether dioecy is more prevalent on islands, previously published accounts of the incidence of dioecy in plant assemblages were collated from the literature. Studies were identified using the search terms 'dioecy', 'sexual system', and 'breeding system' in Google Scholar. To be included, studies had to meet four criteria. They needed to report: (1) the total number of dioecious species in a clearly defined study area, (2) the total number of all species in the study area, (3) the precise location and spatial extent of the study area, and (4) whether sampling was restricted to particular plant life forms.

Recent studies differed widely in the types of plants they considered. Some considered all life forms (e.g., herbs, shrubs, and trees), while others focused on only woody plants or just trees. Studies also differed widely in spatial extent. Many were field surveys conducted on relatively small spatial scales, while others were flora surveys that spanned much larger areas (e.g., New Zealand). To account for the potential effects of sampling area on global patterns in dioecy, studies were classified according to their spatial extent. Studies that spanned more than 100 km^2 were classified as 'coarse scale', while studies that spanned less than 100 km^2 were classified as 'fine scale'.

To evaluate whether global patterns in dioecy were associated with basic climatic variables, temperature and precipitation data for each site were obtained using arcMAP (arcGIS) software to extract values from the WorldClim global dataset (Hijmans et al. 2008) at a 30 arc second resolution (i.e., ~ 1 km^2). Fine-scale studies were only considered if they reported precise geographic coordinates (i.e., latitude and longitude) of their study site. For coarse-scale studies, midpoint values of longitude and latitude across the entire spatial extent of the study region were used.

A linear mixed model was then used to test if dioecy is higher on islands and whether it covaries with global climate. The proportion of dioecious species in each census was used as the dependent variable. Insularity was included as a fixed factor with two levels ('continents' or 'islands'). Precipitation, temperature, and latitude were entered as covariates. In addition to the independent effects of each predictor variable, interactions between temperature, precipitation, latitude, and insularity were also

BOX 4.1 **(cont.)**

assessed. Life form was treated as a random effect with three levels (all growth forms, woody species, or trees). The spatial extent of sampling was included as a second random effect with two levels (coarse scale >100 km^2 > fine scale). Analyses were conducted using the lmer package (Kuznetsova et al. 2017) in the R environment (R Core Team 2013).

One hundred and two published inventories met the inclusion criteria (see Supplementary Resources 4.1). Sixty-one sites occurred on continents and 41 came from islands. The majority of sites (n = 74) reported the incidence of dioecy across all plant species, regardless of growth form, while 17 sites surveyed only woody species, and 11 sites surveyed only tree species. The total pool of data also exhibited considerable variation in the spatial extent of sampling. Fifty-nine sites covered a small spatial extent (fine scale, <100 km^2, n = 18 for islands, n = 41 for continents), while 43 sites covered a large spatial extent (coarse scale, >100 km^2, n = 23 for islands, n = 20 for continents).

Across all categories of life form and sampling area, the proportion of dioecious species ranged between 0 and 26% for islands and between 2% and 40% for continents. Results from the linear mixed model showed the proportion of dioecious species was not statistically associated with insularity, latitude, or temperature (Table B4.1). However, dioecy increased with precipitation (Fig. B4.1). Data courtesy of Matthew Biddick.

Table B4.1 *Results of a linear mixed model investigating the effect of latitude (absolute value), temperature (˚C), precipitation (mm per year), and insularity (island, continent) on the fraction of dioecious species in 102 localities across the globe*

Effect	df	t	p
Insularity	92.140	−1.369	0.175
Precipitation	89.906	4.099	<0.001
Temperature	92.047	−0.527	0.599
Latitude	92.122	−0.183	0.855
Precipitation:insularity	92.129	−1.026	0.308
Temperature:insularity	91.812	1.963	0.053
Latitude:insularity	91.927	1.641	0.104

BOX 4.1   **(cont.)**

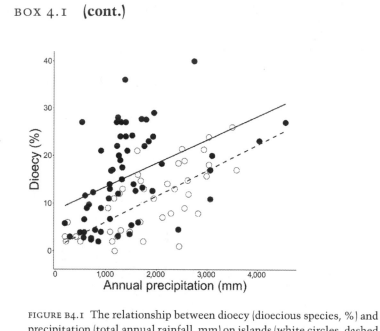

FIGURE B4.1 The relationship between dioecy (dioecious species, %) and precipitation (total annual rainfall, mm) on islands (white circles, dashed line) and continents (black circles, solid line) in 102 plant communities distributed globally.

environments (Herrera 2002a and references within). As a result, interrelationships between outcrossing, seed size, dispersal mode, and precipitation can account for the global distribution of dioecy. This circular relationship between cause and effect is not tautological, however, because it can easily be falsified by disproving any one of the constituent arrows (Fig. 4.3).

Regardless of why dioecy is more prevalent in wet environments, it does not appear to be a hallmark of island floras. However, both this analysis and those of Grossenbacher et al. (2017) leave Baker's (1967) more detailed, time-dependent prediction untested. Gender dimorphism is known to have evolved autochthonously on many isolated islands. For example, 12 lineages of Hawai'ian plants evolved to become gender dimorphic after colonising the archipelago (Sakai et al. 1995). However, the incidence of *in situ* evolution of

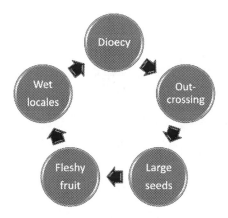

FIGURE 4.3 A mechanistic hypothesis for the relationship between dioecy and precipitation. It predicts that dioecy promotes outcrossing, which promotes larger seed sizes. Larger seeds, in turn, tend to be fleshy-fruited and animal dispersed. Because the incidence of fleshy-fruitedness increases with precipitation across the globe, global patterns in dioecy are predicted to result from cascading effects of dioecy on outcrossing, seed size, dispersal mode, and global climate.

gender dimorphic mating systems on other islands across the globe has yet to be completely resolved. Further study is needed to test whether immigrant selection favours self-compatible island colonists, followed by natural selection for outcrossing mechanisms.

Heterostyly

Heterostyly is a polymorphic mating system that has evolved independently in at least 28 plant families (see Barrett & Shore 2008; Kreiner 2017). Heterostylous species typically produce two flower types, one with long stigmas and short anthers, and the other with short stigmas and long anthers, although tristylous species also exist (Fig. 4.4). These reciprocal differences in flower morphology between morphs facilitates inter-morph pollination by mirroring the positioning of male and female plant organs. Each morph also tends to be self-incompatible, so successful reproduction occurs more frequently between morphs. As a result, heterostyly can help facilitate outcrossing.

In a recent review, Watanabe and Sugawara (2015) tested whether heterostyly is under-represented on islands – a long-standing hypothesis in island biology that is conceptually linked to Baker's law (see Pailler et al. 1998; Watanabe et al. 2014). In a thorough review of previous work, they found that heterostyly is indeed rare on islands. In fact, it appears to be absent in Hawai'i and New Zealand. However, heterostylous taxa

FIGURE 4.4 Heterostylous cowslip (*Primula veris*: Primulaceae) flowers, habit (above) and longitudinal section (below). The long-styled, or 'pin', morph is shown on the left, which has a long style and anthers held low in the corolla tube. The short-styled, or thrum, morph is shown on the right, which has short styles and anthers held at the corolla throat. All are bisexual, but the stigmas recognise and reject pollen from the same morph. Photo taken by Phil Garnock-Jones

occur on other isolated islands, including *Psychotria* spp: Rubiaceae on the Ogasawara Islands, *Cordia lutea*: Boraginaceae in the Galapagos, *Erythroxylum* spp: Erythroxylaceae in the Mascarenes, and *Jasminum odoratissimum*: Oleaceae in the Canary Islands.

Heterostyly is often difficult to recognise in the field and may regularly go unnoticed. Some of the examples listed by Watanabe and Sugawara (2015) were only discovered recently. Therefore, our understanding of the global distribution of heterostyly, both geographically and among plant lineages, is likely to be incomplete. What's more, close examination of some heterostylous species on islands (e.g., *Waltheria ovata*: Malvaceae in the Galapagos) has shown that within morph reproduction can occur sometimes more frequently in island populations than in mainland populations (Bramow et al. 2013).

Anemophily

Most angiosperms are pollinated by animals. In fact, the evolution of animal-pollinated flowers is thought to be a key innovation facilitating their evolutionary diversification. Wind pollination, or anemophily, is therefore a derived condition in many flowering plants. It is also comparatively rare – approximately 10% of angiosperms release their pollen into the atmosphere and utilise air currents as a pollen vector.

Given the importance of animal pollinators in the evolutionary history of angiosperms, why would wind pollination evolve secondarily? One explanation is that anemophily provides reproductive assurance when pollinators are in limited supply (see Friedman & Barrett 2009). Given the depauperate nature of island faunas, islands might exemplify the conditions under which anemophily might be favoured.

In a detailed study of 25 plant species endemic to the Juan Fernández Islands, Anderson et al. (2001) suggest that anemophily was the predominant mode of pollination (see also Anderson & Bernardello 2018). What's more, many species evolved to be wind pollinated autochthonously after colonising the archipelago. *In situ* evolution of anemophily has also occurred elsewhere (e.g., the genus *Schiedea*: Caryophyllaceae in Hawai'i, Golonka et al. 2005; Weller et al. 2006). Most anemophilous species on the Juan Fernández Islands display a suite of morphological characteristics that are typical of wind pollination, including small flower size, reduced pigmentation, small corollas, and long, flexible filaments. However, links between anemophily and particular flower characteristics were more obscure in other species, with some wind-dispersed species having characteristics that are typically associated with animal pollination.

Other studies have failed to find hard boundaries between wind- and animal-pollinated flowers. Travaset and Navaro (2018) identified 11 taxa on Mallorca and other islands in the Western Mediterranean that appeared to be 'ambophilous', or pollinated by both wind and insects. Generalised flower morphologies that can attract generalised pollinators when they are present, and be pollinated by wind when they are absent, may provide further reproductive assurance in island populations.

Is wind pollination unusually common on islands? Although anemophily appears to be common in some archipelagos, work on other islands has found a similar incidence of anemophily to the mainland. For example, many islands in the Mediterranean have reasonably diverse pollinator assemblages and similar numbers of wind-pollinated plants as continental Europe (Traveset & Navarro 2018). Wind pollination on islands in the sub-Antarctic is also similarly common to adjacent continental landmasses (i.e., Patagonia and Tasmania, Lord 2015).

In an ambitious study, Rech et al. (2016) explored global patterns in the incidence of plant species with flower traits that were characteristic of anemophily. They found that wind pollination varied latitudinally and was clearly associated with several climatic variables. Greater numbers of anemophilous species tend to occur in cooler, drier areas, while animal pollination was more prevalent in warmer, wetter areas. After correcting for environmental variation, animal pollination showed a weak decline with insularity, suggesting that wind pollination may indeed be a general characteristic of island floras. However, only 11 of their 67 study sites were islands.

Generalised Flower Form

Pollinators that are common on continents often fail to disperse to isolated islands. As a result, assemblages of pollinators on islands often differ substantially from mainland assemblages. These changes could have profound implications for animal-pollinated plants, particularly those that have evolved specialised mutualistic relationships with particular types of pollinators. When a plant species with a specialised flower form disperses to an island, chances are it will do so without its pollinator partner. Under these circumstances, more generalised relationships with animal pollinators may be advantageous, as flowers with a more generalised morphology are more likely to attract a suitable pollinator from within depauperate pollinator faunas (Fig. 4.5).

(a)

FIGURE 4.5 Examples of small insular flowers (a) *Melicytus novae-zelandiae*: Violaceae, whose flowers are 3–4 mm in diameter, (b) *Kunzea ericoides*: Myrtaceae, whose flowers are <1 cm in diameter, which is smaller than most congeners in Australia (photo taken by Phil Garnock-Jones), and (c) male flowers of *Pseudopanax arboreus*: Araliaceae, whose flowers are <1 cm in diameter.

(b)

(c)

Many studies have tested whether the flowers produced by island plants are more generalised than mainland plants. In a pioneering study, Spears (1987) compared the pollination biology of *Centrosema virginianum*: Fabaceae, which produces specialised flowers, and *Opuntia stricta*: Cactaceae, which produces more generalised flowers, between Florida and several adjacent islands in the Caribbean Sea. Results showed that pollen dispersal to neighbouring plants was lower in island populations of *C. virginianum*, consistent with the depauperate nature of pollinator faunas. A similar, yet weaker, pattern was observed in *O. stricta*. Similarly, Martén-Rodríguez et al. (2015) found that pollination processes across the Caribbean in the family Gesneriaceae tend to be more generalised than on the mainland.

A series of related studies on islands located off the coast of Japan have explored relationships between the morphology of flowers and their pollinators. Inoue and Amano (1986) found that islands tended to lack large-bodied pollinators (e.g., bees), and, in their absence, populations of *Campanula punctata*: Campanulaceae produced smaller flowers. Yamada et al. (2014) found similar results in *Ligustrum ovalifolium*: Oleaceae. Concomitant with reductions in the diversity of pollinators, island plants produced smaller stamens and corolla tubes than populations on the Japanese mainland. Additional comparisons of corolla tube lengths between island and mainland subspecies of *Weigela coraeensis*: Caprifoliaceae showed that they declined on more isolated islands (Yamada et al. 2010). Similarly, Yamada and Maki (2014) found that the corolla tubes of *Hosta longipes*: Asparagaceae were shorter on the more isolated Izu Islands, suggesting a trend towards more generalised flower morphology.

Schueller (2007) compared the floral morphology of *Nicotiniana glauca*: Solanaceae between the California Islands and the adjacent mainland. Island populations were visited by longer-billed hummingbirds than on the mainland. In an apparent trend towards increased specialisation, they also had longer corollas. However, more detailed analyses of pollinator-mediated selection pressures showed that longer-billed hummingbirds were not selecting for longer-corollaed flowers during the time frame of the study.

Islands in the Mediterranean may be an exception to a general trend towards increased floral generalisation on islands (Lázaro & Traveset 2005). On many Mediterranean islands, pollinator diversity is not conspicuously lower than it is on the mainland. Furthermore, many islands in the Mediterranean support plant taxa with highly specialised flowers that are pollinated by one or a small number of pollinator species (see Traveset & Navaro 2018 for a recent review).

Although these studies suggest that insular selection pressures may favour increased generalisation of flowers on islands, whether flower generalisation is a repeated pattern in island evolution is difficult to pinpoint, in part because flower morphology is difficult to characterise with one or a small number of metrics. This severely hampers efforts to test for overarching trends towards flower generalisation among archipelagos. Overcoming this hurdle towards a better understanding of flower evolution on islands presents a serious challenge to future research.

Reduced Floral Diversity

Reductions in pollinator diversity on islands may have important implications for the diversity of flower sizes produced by island plants. As pollinator diversity declines, so too should the overall size range of pollinator species. Size coupling is an important aspect of pollination mutualisms (see Biddick & Burns 2018). Smaller flowers tend to be pollinated by smaller pollinators, while larger flowers tend to be pollinated by larger pollinators. Therefore, island plant communities should exhibit reduced flower-size diversity relative to continental floras.

To test this prediction, the size of flowers produced by 11 species of plants endemic to small islands in the South-west Pacific were measured and compared to the size of flowers produced by their closest relatives on the New Zealand 'mainland' (Box 4.2). Results

BOX 4.2   **Reduced flower-size diversity in the South-west Pacific**

Species diversity in most types of organisms declines with insularity. Flower-visiting insects are no exception (see Barrett 1996). Assuming different-sized pollinators select for different-sized flowers (see Biddick & Burns 2018), flower-size diversity should decline on islands. To test this hypothesis, insular changes in flower size were assessed in 11 species pairs of animal-pollinated plants in the South-west Pacific.

All species listed in Poole and Adams (1994) as endemic to small islands surrounding New Zealand were considered, as well as their closest relative on the New Zealand 'mainland', which was determined using molecular tools (Supplementary Resources 4.2). The size of flowers produced by each taxon was obtained from taxonomic descriptions in the New Zealand flora (Allan 1961). Species listed by Poole and Adams (1994) were omitted if consistent descriptions of flower sizes were unavailable, the closest mainland relative could not be identified or if 'island' species maintained populations in New Zealand as well as offshore islands. Species arising from cladogenesis on islands were also not considered.

To test whether flower-size diversity differed between island taxa and their mainland relatives, flower sizes were compared with reduced major axis regression using the lmodel2 package in the R environment (R Core Team 2013). A slope parameter less than one and an intercept parameter greater than zero would indicate flower size diversity is lower on islands (see Box 1.5). To further test for insular size changes in flower size, an additional test was conducted. Insular size changes were calculated by dividing island flower sizes by mainland flower sizes. Insular size changes were then regressed against mainland flower sizes. A negative relationship would indicate that smaller flowers evolve to become larger on islands, while larger flowers evolve to become smaller, resulting in lower flower-size diversity.

BOX 4.2   **(cont.)**

Results showed that island flower sizes were less diverse than
mainland flower sizes (Fig. B4.2). Reduced major axis regression
produced a slope parameter than was less than one (estimate = 0.638,
95% confidence interval = 0.405–0.934) and an intercept
parameter that was greater than zero (estimate = 0.762,
95% confidence interval = 0.230–1.182). Insular size changes were also
negatively related to mainland flower sizes (t = –3.801, p = 0.004).

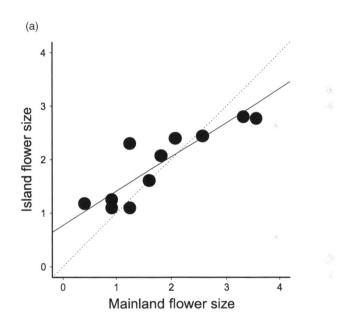

(a)

FIGURE B4.2 (a) Scaling relationship between the size of flowers
produced by 11 island taxa and their closest mainland ancestors. The
relationship differed from isometry (slope <1, intercept >0), indicating
flower-size diversity is lower in island taxa. (b) Negative relationship
between insular size changes (island flower sizes ÷ mainland flower
sizes) and mainland flower sizes in the same data, indicating smaller-
flowered species evolve to become larger on islands, while larger-
flowered species become smaller.

BOX 4.2   **(cont.)**

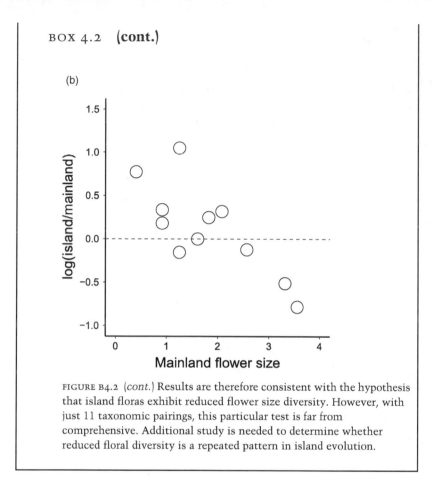

FIGURE B4.2 (*cont.*) Results are therefore consistent with the hypothesis that island floras exhibit reduced flower size diversity. However, with just 11 taxonomic pairings, this particular test is far from comprehensive. Additional study is needed to determine whether reduced floral diversity is a repeated pattern in island evolution.

showed that small-flowered species tended to become bigger on islands, while large-flowered species tended to become smaller, resulting in reduced flower-size diversity.

CONCLUSIONS

Unlike other aspects of the island syndrome in plants, the reproductive biology of island plants has been studied extensively for decades. The rather large body of empirical work on the subject has also been thoroughly reviewed at various times in the past (e.g., Barrett 1996; Crawford et al. 2011; Anderson and Bernardello 2018; Traveset and Navaro 2018). So, instead of providing an exhaustive review of the reproductive biology of island plants, this chapter briefly summarised the evidence for several traditional lines of research. It also explored several under-studied aspects of gender and outcrossing in island plants.

Most evidence to date is consistent with Baker's law. Plants that are capable of uniparental reproduction are unusually common in island floras. One potentially confounding effect on the results of previous tests of Baker's law is the time since population establishment. Baker's law predicts that immigrant selection favours self-compatibility only at the initial stage of population establishment. Following population establishment, self-compatibility may quickly become disadvantageous, and selection could then favour self-incompatibility (Baker 1967). If these two processes operate simultaneously, it would lead to the co-occurrence of young, self-compatible populations, as well as older island populations that have evolved *in situ* to become self-incompatible. An important challenge for future research will be to disentangle the effects of these two opposing processes on the expression of self-compatibility.

Contrary to earlier speculation, global-scale analyses of the incidence of dioecy indicates that it is not higher on islands than it is on continents. A probable cause for the hypothesised link between dioecy and insularity is that it arose as a by-product of a chain of relationships among outcrossing, seed size, fleshy-fruitedness, and precipitation. On the other hand, heterostyly appears to be under-represented on islands.

Anemophily may very well be a component of an island syndrome in plants. However, many unanswered questions about its incidence in island floras remain. How regularly do floral traits accurately predict anemophily? As the incidence of anemophily is established across a greater number of archipelagos, does it continue to characterise islands in general? If so, does insular anemophily result from pollinator limitation on islands or does it arise because islands tend to be windier places that continents? How frequently does anemophily arise on islands via immigrant selection and how often might it evolve autochthonously?

Many previous studies have shown that evolution tends to favour more generalised flowers on islands. These studies suggest that in the absence of specialised pollinators specialised structures associated with them are selected against. Although numerous studies on specific species have shown this to be the case, flower 'generalisation' is difficult to characterise uniformly across taxa, and an overarching test for repeated patterns in flower generalisation has yet to be conducted. One way forward is to focus on flower size and test whether island flowers are consistently smaller than mainland flowers. Analyses of several species on small islands in the South-west Pacific suggest that this might be a useful way forward.

At present, only Baker's law can safely be considered a component of the island syndrome in plants that is related to reproductive biology. However, several other aspects of plant reproductive biology appear to be shaped consistently by insular selection pressures, but require further study. Anemophily, reduced flower-size diversity and the increased incidence of outcrossing mechanisms such as heterostyly in more derived plant lineages may also have evolved repeatedly on islands.

5   Size Changes

The life history of animals is heavily influenced by their body size. Bigger animals live longer, move more slowly, and reproduce less frequently than smaller animals. Given its functional significance, understanding how selection shapes the body size of animals is an important goal in organismal biology.

Interestingly, animals that are endemic to isolated islands often have very different body sizes than their relatives on the mainland. Paradoxically, some animals evolve to become bigger on islands, while others evolve to become smaller (Foster 1964). Elephant birds and giant lemurs shared their island home with pigmy hippos and dwarf chameleons. This paradox of island evolution has fascinated biologists for centuries and in many ways remains an enigma.

Many studies have tested for a graded trend in body-size evolution on islands, wherein small animals evolve to become larger and large animals evolve to become smaller (e.g., Lomolino 2005; Lomolino et al. 2006; Meiri et al. 2006, 2008; Faurby & Svenning 2016). Results have varied among taxa and between research groups (Lokatis & Jeschke 2018), leading to a long history of controversy and debate. Nevertheless, the island rule has been an invaluable paradigm for zoologists. By making basic predictions that can be tested with simple estimates of body mass, the island rule has unified research into the evolution of body size in island animals.

The history of scientific inquiry into size changes in island plants has taken a very different path. Over 150 years ago, Darwin (1859) noticed that the closest relatives of many island tree species were small-statured herbs on continents (Fig. 5.1). This led him to speculate that selection favours increased stature in island plants and ultimately increased woodiness. Darwin's 'weeds-to-trees' hypothesis

FIGURE 5.1 *Lordhowea insularis*: Asteraceae, an example of insular woodiness. It is a woody perennial that is endemic to Lord Howe Island and grows to 2 m in height. Molecular phylogenetic analyses identified *Phaneroglossa bolusii*, a herbaceous annual plant from Southern Africa that grows to 0.6 m in height, as its closest relative (Pelser et al. 2002).

left an indelible impression on botanists, who have focused on insular woodiness ever since (e.g., Carlquist 2001).

There are a number of reasons why the search for an island syndrome in plants should expand to encompass the island rule, rather than focus solely on insular woodiness. First, plants can vary in 'size' in a number of interesting and important ways without concomitant changes in woodiness. Most notably, they can produce different-sized seeds or leaves. A focus solely on insular woodiness will leave the evolutionary dynamics of these traits unexplored.

Second, different plant traits may evolve cohesively, making it difficult to pinpoint which trait might be under selection. For example, if selection favours larger leaves, woody stems might also evolve to provide structural support. Webs of genetic and developmental associations can also lead to correlated evolution of apparently unrelated plant organs, such as leaves and fruits (Box 1.6). This perspective opens the door to the possibility that the evolution of woodiness could arise as a passive by-product of selection on other plant traits, such as plant stature.

Lastly, remarkable changes in plant size occur on islands that do not support woody plants. For example, many plants inhabiting high-latitude islands just north of Antarctica (e.g., Campbell and Macquarie Islands) exhibit dramatic changes in size relative to their mainland ancestors. However, the climatic conditions on sub-Antarctic islands prohibit the proliferation of trees and shrubs. Instead, a diverse array of herbaceous plants has evolved into 'mega-herbs' on sub-Antarctic islands (Fig. 5.2), for reasons that are, for the most part, completely decoupled from woodiness *per se*.

SYNDROME PREDICTIONS

This chapter explores the rich diversity of size changes exhibited by island plants. First, it briefly summarises our current understanding of the evolution of insular woodiness. It then moves on to explore changes in plant stature, leaf area, and seed size. Previous work on each of these traits is far more limited than previous work on insular woodiness. To bridge this gap in our understanding of island plants, new data are used to test for repeated patterns in island evolution. More specifically, these data are used to provide the first test of the island rule in plants.

FIGURE 5.2 Subantarctic 'megaherbs' near Col-Lyall Saddle, Campbell Island. Prominent are *Pleurophyllum speciosum:* Asteraceae with disk-shaped flowers (c. 60 mm diameter) and *Bulbinella rossii:* Asphodelaceae (erect racemes stretching to 60 cm long). (photo taken by Phil Garnock-Jones)

HYPOTHESIS TESING

Insular Woodiness

The life history of angiosperms varies enormously, from tiny herbs that live for just a few weeks, to tall trees that live for centuries. Woodiness is a key structural trait facilitating the evolution of large, perennial growth forms. Because angiosperms likely evolved from gymnosperm-like ancestors, woodiness is often considered an ancestral trait in flowering plants (see Dulin & Kirchoff 2010). Conversely, herbaceousness is thought to be a derived condition. When woodiness evolves in herbaceous plants, it is known as *secondary woodiness*.

Darwin (1859) was among the first to notice that the closest relatives of many island tree species are continental herbs. When a lineage of predominately herbaceous species has woody representatives on islands, it is known as *insular woodiness* (Dulin & Kirchoff 2010). It is commonly assumed that insular woodiness arises from *in situ* selection for woodiness on islands. However, this might not always be the case. Secondary woodiness often evolves in continental plant taxa, which later colonise islands. If the lineages to which these species belong later go extinct on the mainland, the presence of secondary woodiness on the island would be relictual (Cronk 1992).

There is evidence for both evolutionary pathways towards woodiness on islands. Molecular phylogenetic analyses of many insular linages, including *Descurainia*: Brassicaceae (Goodson et al. 2006), *Pericallis*: Asteraceae (Swenson & Manns 2003), and *Tolpis*: Asteraceae (Moore et al. 2002) illustrate that woodiness is relictual. However, an apparently larger body of evidence indicates that insular woodiness is a derived condition, including *Sideritis*: Laminaceae (Barber et al. 2002), *Echium*: Boraginaceae (Böhle et al. 1996), and *Lavatera*: Malvaceae (Fuertes-Aguilar et al. 2002) in Macaronesia, the Hawai'ian silverswords (*Argyroxiphium*, *Wilkesia*, *Dubautia*: Asteraceae), and *Sanctambrosia manicata*: Caryophyllaceae (Kool & Thulin 2017) on San Ambrosio Island, off the coast of Chile.

Lens et al. (2013) reviewed the evidence for insular woodiness in the flora of the Canary Islands. They found that insular woodiness (i.e., *in situ* evolution of secondary woodiness) was exceedingly common (Fig. 5.3), having occurred on 38 occasions, leading to a total of 220 species from 34 genera that exhibit insular woodiness. This suggests that insular woodiness is a repeated pattern in evolution on isolated islands, and thus a component of an island syndrome in plants. However, Lens et al. (2013) argue that more extensive analyses of the evolution of secondary woodiness across the globe is needed before reaching this conclusion. Secondary woodiness has evolved repeatedly on continents (>200 genera), and it has yet to be demonstrated quantitatively that *in situ* evolution of woodiness on islands is

FIGURE 5.3 *Sonchus hierrensis*: Asteraceae, an example of insular woodiness from Isla La Palma, Canary Islands.

more common than relictual woodiness (Whittaker & Fernández-Palacios 2007).

Why might increased woodiness be adaptive on islands? Charles Darwin, Alfred Russel Wallace, and Sherwin Carlquist all derived unique hypotheses to explain insular woodiness (see Whittaker & Fernández-Palacios 2007). Darwin's (1859) *weeds-to-trees hypothesis* argues that when islands are young geologically, they are more likely to be colonised by weedy, herbaceous species because herbs tend to have smaller, better-dispersed seeds than trees. After island colonisation, herbaceous colonists then evolve a taller stature and increased woodiness through time to improve their capacity to compete for light, eventually filling the empty niche occupied by trees on continents. Darwin's hypothesis therefore implies that selection favours woodiness as a by-product of selection for taller stature.

Wallace's *lifespan hypothesis* argues that woodiness evolves on islands to increase plant longevity, and longer individual lifespans are selectively advantageous because they lead to longer flowering times. Longer flowering times are particularly advantageous in insular environments, given the depauperate nature of island pollinator faunas, because longer flowering times will increase the probability they will be visited by pollinators, all else being equal (see Givnish 1998).

Carlquist (1974) suggested that insular woodiness arises from the stability of insular climates. His *insular climate hypothesis* assumes that herbaceousness is an effective way to avoid periods of harsh climatic conditions on contents, such as freezing temperatures in polar latitudes, or extended periods of drought in savannas or deserts. Given the mild, stable climates of oceanic islands, the ability to persist through unfavourable climates as seeds may no longer be advantageous, thus allowing plants to become perennial and ultimately woody.

An alternative explanation for insular woodiness is that it results passively via correlated evolution with leaf size (Burns 2016b). A variety of processes may select for large leaves on islands. For example, large leaves may perform better physiologically in wetter island environments (Wright et al. 2017). Alternatively, small leaf sizes are known to deter a wide range of herbivores (see Hanley et al. 2007). On islands that lack these herbivores, plants may be released from selection for microphylly and begin to produce large leaves. Given that larger leaves are heavier, plants may in turn evolve woody stems to provide mechanical support.

Stature

Plants on isolated islands often evolve dramatic differences in stature in short periods of evolutionary time. Molecular phylogenetic analyses of plant lineages undergoing adaptive radiation illustrate that island selection pressures can generate tremendous morphological diversity (Givnish 1997; 1998; 2015). For example, there are approximately 37 species in the genus *Echium*: Boraginaceae on islands in Macronesia, all of which evolved from a single herbaceous ancestor that dispersed to the islands approximately 20 Ma (Böhle et al. 1996). The lineage has subsequently radiated and the genus now contains a remarkable diversity of growth forms, including small-statured annuals, perennial rosettes, shrubs, and even trees.

The Hawai'ian silversword alliance is comprised of approximately 30 species that evolved from a herbaceous ancestor that arrived in the Hawai'ian archipelago approximately 5 Ma (Baldwin & Sanderson 1998). It contains three genera (*Argyroxiphium*, *Dubautia*, *Wilkesia*: Asteraceae), which encompass an even greater range of growth forms than *Echium* in Macaronesia, including cushion plants and vines (Robichaux et al. 1990). A parallel, yet less diverse radiation of silverswords also occurred on the California Islands (Baldwin 2007).

The largest plant radiation on the Hawai'ian Islands are the lobeliads (Campanulaceae), which encompass approximately 125 species, all of which evolved from a single common ancestor that dispersed to the archipelago approximately 13 Ma (Givnish et al. 2009). A variety of growth forms are again represented, from short-statured rosettes to tall trees, which display a diverse array of physiological adaptations to different environmental conditions (Montgomery & Givnish 2008). Yet unlike the Hawai'ian silversword alliance and Macronesian *Echium*, the Hawai'ian lobeliads evolved from a woody rather than herbaceous ancestor.

The genus *Erigeron* represents the best example of adaptive radiation on the Juan Fernandez islands, which are located approximated 700 km off the coast of South America in the Southern Pacific Ocean (Stuessy et al. 2018). Six species are currently recognised, all of which evolved via cladogenesis from a single ancestor (Valdebenito et al. 1992). The genus exhibits a range of growth forms, from short-statured sub-shrubs to taller woody forms (see also Takayama et al. 2018).

On the other side of the Pacific Ocean from the Juan Fernandez Islands lies Lord Howe Island, where many species have evolved via anagenesis (Papadopulos 2011). However, other endemic species have evolved sympatrically, via cladogenesis. Arguably the best known example is two species of palms in the genus *Howea*: Arecaceae, which evolved from a single colonisation event, but now markedly differ in stature, distribution and flowering phenology (Savolainen et al. 2006).

These and other phylogenetic analyses of plant lineages that have evolved via cladogenesis clearly indicate that island selection pressures can favour pronounced differences in plant morphology, often over relatively short periods of time. However, much less is known about repeated patterns in morphological evolution of island species that evolved anagenetically. A significant fraction of island endemics have speciated alone, rather than radiated rapidly (see Takayama et al. 2018). The products of anagenetic evolution are especially abundant on low-lying islands containing a low diversity of habitats, where anagenesis is the predominant mode of evolution, not cladogenesis (Stuessy et al. 2006). Pronounced changes in stature can also take place in the absence of speciation. However, repeated patterns in the morphological evolution of populations that have yet to speciate on islands are also poorly resolved.

Hochberg (1980) provides a notable exception. He compared the morphology of three species of chaparral shrubs between the California Islands and the nearby mainland. *Prunus ilicifolia*: Rosaceae was less branched close to the ground, indicating that it tends to grow from a single central axis (i.e., trunk) on islands, rendering it more arborescent. However, the other two shrub species (*Ceanothus megacarpus*: Rhamnaceae; *Dendromecon rigida*: Papaveraceae) showed the opposite pattern. Insular populations produced greater numbers of branches at ground level, suggesting that they had evolved to become shrubbier.

Burns (2016b) compared the size of *Alyxia ruscifolia*: Apocynaceae between Lord Howe Island and Southern Queensland, Australia. Reproductively mature plants on the island were shorter in stature than mature plants on the mainland. Island plants also matured at shorter heights than mainland plants. Results therefore showed clear evidence for dwarfism in plant stature; however, these results are hardly generalisable, as only two populations of a single species were considered.

Pandanus forsteri: Panadaceae is a species of screw pine this is endemic to Lord Howe Island. Its closest relative is *Pandanus kanehirae*, which occurs in South-east Asia (Gallagher et al. 2015). *P. forsteri* is much larger than *P. kanehirae*, providing an example of insular gigantism. However, *P. forsteri* is a monocot and cannot add additional support radially as it grows taller. Instead, it supports its massive weight with a lattice of lateral proproots. This creates a problem, however, as aerial roots do not have access to water initially as they grow. *P. forsteri* has solved this problem by capturing water with gutter-shaped leaves, which is then transported via gravity down channels located on its stems towards aerial root tips. Aerial root tips are also covered in a distinctive fibrous tissue that stores water, keeping root tips moist as they grow towards the soil surface (Biddick et al. 2018).

Burns et al. (2012) compared the heights of six species of *Veronica*: Plantaginaceae that are endemic to small islands offshore of New Zealand to their relatives on the New Zealand 'mainland'. Some species increased in size (e.g., *Veronica barkeri*), while others declined (e.g., *Veronica chathamica*). A preliminary test of the island rule showed that although the slope parameter of the relationship between island stature and mainland stature was less than one, it was not statistically significant.

In a more comprehensive study, Cox and Burns (2017) compared the stature of dozens of plant taxa on the Chatham Islands with their closest relatives growing on larger landmasses. They measured the height of sexually mature plants of nine species pairs by hand in the field and obtained height estimates for 30 endemic taxa from the literature. Taxonomic pairings between Chatham Island species and their closest mainland relatives were established using molecular tools (Heenan et al. 2010). Results showed that some island plants increased stature on the Chatham Islands, while others decreased. However, there was weak overall trend towards gigantism.

Box 5.1 provides the first thorough test for the island rule in the stature of island plants, using data from Cox and Burns (2017). Consistent with the island rule, results illustrate a graded trend in plant size, with the directionality of size changes being related to plant size itself. Smaller-statured species tend to increase in stature on the Chatham Islands, while larger-statured species tend to decrease in stature.

What might select for the island rule in plants? The occurrence of both dwarfism and gigantism in plant stature is inconsistent with Darwin's (1859) competition hypothesis, which predicts plants undergo directional selection for gigantism, along a weeds-to-trees evolutionary pathway. No previous study has tested for differences in longevity of island and mainland populations, so Wallace's (1878) longevity hypothesis remains untested. Most of the species in the study are woody perennials, so Carlquist's climatic stability hypothesis does not appear to apply.

Lomolino (2005) hypothesised that the island rule results from relaxed predation pressures and reduced competition on islands. He reasoned that because continents tend to house greater diversities of predators and competitors, selection favours greater phenotypic divergence in mainland species to avoid predators and mediate interspecific competition. However, when released from divergent selection pressures in depauperate insular environments, small island species evolve to become larger and large island species evolve to become smaller. Changes in the stature of island plants depicted in Box 5.1 are strikingly consistent with Lomolino's (2005) explanatory model for the island rule (Fig. 5.4).

A variety of mechanisms can promote competitive coexistence among species in plant communities (see Chase & Leibold 2003). For instance, plants may be differentially adapted to exploit distinct periods of time following disturbance, and thereby fill different 'regeneration niches' (Grubb 1977). Some species may be small in stature and have quick generation times to exploit recently disturbed areas. At the other end of the spectrum, taller-statured

BOX 5.1   **Stature of Chatham Island plants**

To investigate consistent evolutionary size changes in the stature of Chatham Island plants, two data sets were used to test the 'island rule' (see Supplementary Resources 5.1 and 5.2). The first is comprised of field measurements of nine taxonomic pairings between species endemic to the Chatham Islands and their closest ancestral species on the New Zealand 'mainland'. The second is comprised of maximum height values for 23 sister species (see Heenen et al. 2010) obtained from the literature. Endemic species that originated via cladogenesis were removed to promote independence among replicates.

Insular size changes were calculated by dividing the average stature of each island taxa by the average stature of their corresponding mainland sister taxon. Insular size changes were log transformed to conform to assumptions, so values greater than zero represent gigantism and values less than zero represent dwarfism. General linear models were then used to test whether insular size changes were associated with the size of mainland taxa. Separate tests were conducted on field- and literature-derived data. Island–mainland size ratios were taken as the dependent variable and the stature of mainland taxa was used as a covariate in both models. Analyses of literature-derived data also included plant growth form (herbaceous versus woody) as a fixed factor with two levels.

Results were consistent with the island rule (Fig. B5.1). Insular size changes were negatively related to mainland stature in both field data ($F = 25.683$, $p = 0.001$) and literature-derived data ($F = 23.444$, $p < 0.001$). In the literature-derived data, mainland stature did not interact with growth form ($F = 0.847$, $p = 0.369$). However, a significant effect of growth form was observed ($F = 27.509$, $p < 0.001$), indicating the y-intercept for herbs was lower than y-intercept for woody plants. This illustrates that while both herbs and woody plants conform to the island rule, herbaceous plants converge on a shorter stature than woody plants, which is consistent with Lomolino's (2005) fundamental size theory.

BOX 5.1 **(cont.)**

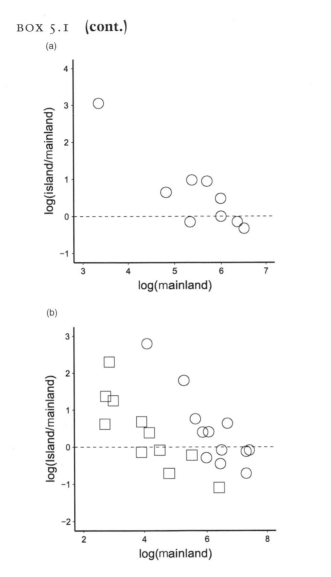

FIGURE B5.1 Evidence for the island rule in the stature of Chatham Islands plants. Island–mainland ratios of plant stature (y-axis) decline with mainland values of stature (x-axis) in both analyses. (a) Averages of field measurements for nine island–mainland species pairs. (b) Maximum height values for 23 taxonomic pairings obtained from the literature (squares: herbaceous species; circles: woody plants).

(a)

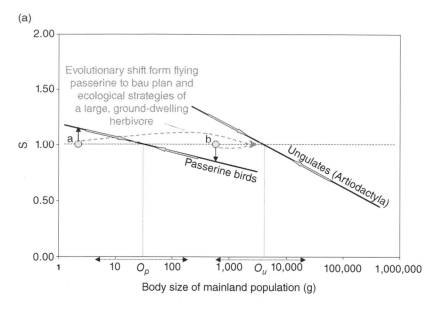

FIGURE 5.4 (a) Lomolino's (2005) explanatory model for the island rule (reproduced with permission). Because islands tend to house fewer species than the mainland, the model assumes that interspecific competition (and other ecological interactions, such as predation) is lower on islands. So, when an organism colonises an island from the mainland, it will be released from selection for phenotypic differentiation to promote coexistence, thus favouring phenotypic convergence. Small animals become bigger on islands, whereas bigger animals become smaller. This is illustrated graphically by plotting the ratio of island body size to mainland body size (y-axis, S_i) against mainland body size (x-axis). The dashed, horizontal line represents no differentiation in body size between islands and the mainland. General trends in two taxa (passerines and ungulates) are illustrated as solid curves with a negative slope. Grey points and arrows (a and b) illustrate gigantism and dwarfism, respectively. O_p and O_u denote fundamental or 'optimal' body sizes for each taxonomic group. (b) Insular changes in the stature of plants inhabiting the Chatham Islands (axes are identical to a, data are identical to Fig. B5.1). Separate best-fit lines are shown for herbs (squares) and woody plants (circles), whose slopes are similar but intercepts differ. The dashed grey line illustrates the weeds-to-trees evolutionary pathway.

species with longer generation times may dominate areas long after they were last disturbed. Assuming landscapes are comprised of a large enough matrix of habitat patches that were disturbed at different times in the past, interspecific differences in regeneration niches

(b)

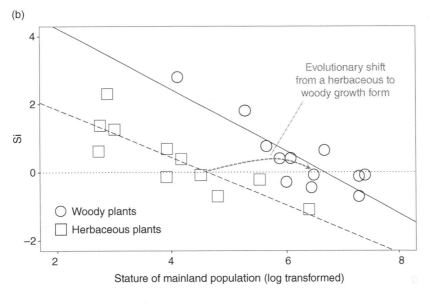

FIGURE 5.4 (cont.)

can promote competitive coexistence in diverse plant communities. However, given their restricted spatial extent, this assumption may often be violated on islands, relaxing selection for divergent regeneration niches and therefore plant stature.

Leaf Area

In addition to testing for differences in plant stature, Hochberg (1980) investigated the size of leaves produced by insular populations of *P. ilicifolia*, *C. megacarpus*, and *D. rigida* on the California Islands. All three species produced much larger leaves on the islands than on the mainland. No evidence for leaf 'dwarfism' was observed.

Beever (1986) described differences in leaf size in a wider range of species inhabiting islets off the North Island of New Zealand. Comparisons between mainland and island populations (n = 13 species, including *Rhabdothamnus solandri*, Gesneriaceae), subspecies (n = 5 comparisons, including *Solanum aviculare* var. *latifolium* vs var. *aviculare*, Solanaceae) and between closely related species (n = 7 comparisons, including *Pennantia baylisiana* vs *Pennantia corymbosa:* Pennantiaceae) all showed consistent increases in leaf size.

Working on other islands in the South-west Pacific, Burns et al. (2012) investigated changes in the size of leaves produced by a large number of species from a wide range of phylogenetic backgrounds (see Fig. 5.5). Despite strong differences in latitude, from Campbell Island in the south (−52°, 54′) to the Kermadecs in the north (−29°, 27′), they too found consistent support for leaf gigantism.

Less clear results were obtained from comparisons of leaf size between populations of *A. ruscifolia* on Lord Howe Island and the Australian mainland (Burns 2016b). Island plants produced leaves with the same average size as plants on the mainland. However, they were much less variable. Plants on the mainland of Australia exhibited stronger ontogenetic variation in leaf area, with shorter plants producing disproportionately smaller, narrower, and more spinescent leaves, and taller plants producing larger leaves that were less spinescent. Therefore, insular changes in leaf size in *A. ruscifolia* were linked to leaf spinescence and presumably plant defence.

FIGURE 5.5 The larger leaf on the right is *Macropiper hooglandii*: Piperaceae, an endemic species of shrub from Lord Howe Island that exemplifies leaf gigantism. On the left is *Macropiper excelsum*, a more recent immigrant from New Zealand.

Additional support for leaf gigantism was found by Cox and Burns (2017). They measured the size of leaves produced by 18 species on the Chatham Islands by hand, and compared them to analogous field measurements on their closest relatives occurring on larger landmasses. They also conducted similar comparisons using literature-derived estimates of maximum leaf area from 30 species. Results showed support for leaf gigantism. Furthermore, the magnitude of leaf size increases was also related to how long each taxon has resided in the Chatham Islands. Insular size changes (i.e., the ratio of average leaf area in island plants to average leaf area in their sister species growing elsewhere) were positively related to divergence times, which were estimated using molecular clock techniques (Heenan et al. 2010). Therefore, taxa that have resided on the Chatham Islands for longer periods showed stronger evidence of leaf gigantism than those that arrived more recently.

Kavanagh and Burns (2015) tested for increased sexual size dimorphism in dioecious plants on isolated islands in the South-west Pacific, as predicted by the niche variation hypothesis (Van Valen 1965). Results were inconsistent with its predictions, as size trends were similar in both males and females; island plants consistently produced larger leaves.

In a similar study focusing on branching architecture, Kavanagh (2015) compared populations of divaricately branched species in New Zealand to populations inhabiting smaller islands offshore. He found that island plants produced shallower branch angles, shorter internodes, and larger leaves. Results were therefore consistent with the hypothesis that when released from vertebrate herbivory on islands devoid of vertebrate herbivores, plants lose many of the characteristics associated with the divaricate growth form, most notably microphylly. Kavanagh's (2015) study therefore suggests that leaf gigantism might be linked to differences in herbivore communities on islands.

Biddick and Burns (2019) analysed insular changes in the sizes of leaves produced by cosmopolitan lianas occurring across the

South-west Pacific. They found that island populations consistently produced larger leaves than those of mainland populations. Leaf gigantism in these species could not be explained by an associated increase in stature or woodiness, given that lianas are structurally parasitic and lack independent stature. Therefore, changes in leaf size do not appear to result from allometric relationships with plant stature. What's more, changes in leaf size were unrelated to changes in stem size. They conclude that leaf and twig size are under disparate selection pressures, and that insular leaf gigantism is not contingent on an associated increase in stature or stem size.

Working in Macaronesia, García-Verdugo et al. (2010a) investigated both morphological and genetic differences among subspecies of *Olea europaea*: Oleaceae in Southern Europe, North Africa, the Canary Islands, and Madeira (see also García-Verdugo et al. 2010b). They found that the different subspecies were strongly subdivided genetically, with the two insular subspecies producing larger leaves than their European counterparts. Interestingly, they also found that leaf mass per unit area differed consistently in island populations, with island plants producing leaves with less mass per unit area than mainland plants. García-Verdugo et al.'s (2010a) results therefore suggest that island plants could show consistent differentiation in traits related to the leaf economic spectrum, in addition to the traits considered here (i.e., woodiness, stature, leaf area, and seed size). Testing for consistent differentiation in the positions of island species along the leaf economics spectrum is likely to be a very interesting avenue of future research (see Westoby 1998; Wright et al. 2004).

A clear consensus emerges from this review of previous work on the sizes of leaves produced by island plants. Leaf-size gigantism is a common pattern in island evolution. Box 5.2 takes one step further and tests whether changes in leaf size in island populations are associated with the size of ancestral mainland populations using all available data from the South-west Pacific. Widespread support for leaf gigantism was again observed. However, results also showed partial support for the island rule. More specifically, the extent of leaf

BOX 5.2 Leaf size in the South-west Pacific

To investigate evolutionary size changes in leaves produced by plants inhabiting small islands surrounding New Zealand (Chatham, Three Kings, and Kermadec Islands), two data sets were used to test the 'island rule' (see Supplementary Resources 5.1 and 5.2). The first is comprised of field measurements of 24 pairings between insular taxa and their closest ancestor on the New Zealand 'mainland' (see Heenen et al. 2010). The second data set is comprised of maximum leaf area estimates obtained from the literature for 23 species pairs inhabiting the Chatham Islands. Endemic species that originated via cladogenesis were removed to promote independence among replicates. Statistical tests of the island rule were conducted as in Box 5.1.

(a)

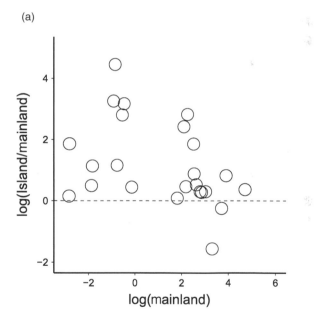

FIGURE B5.2 Evidence for gigantism and partial support for the island rule in leaf sizes of plants inhabiting small islands surrounding New Zealand. Island-mainland ratios of leaf size (y-axis) are plotted against mainland values (x-axis) in both analyses. (a) Averages of field measurements for 24 island-mainland species pairs. (b) Leaf size estimates for 23 taxonomic pairings obtained from the literature (squares: herbaceous species; circles: woody plants).

BOX 5.2 **(cont.)**

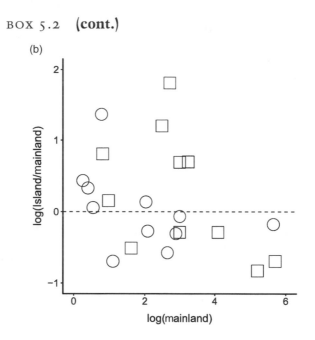

FIGURE B5.2 (*cont.*)

Results from field data showed strong support for leaf gigantism and weak support for the island rule (Fig. B5.2). Field data showed that 92% of taxonomic pairings (22/24 species pairs) increased in size on islands, insular increases in leaf size were also weakly associated with the size of leaves on the mainland (F = 4.363, p = 0.049). Therefore, smaller-leaved species tended to show stronger increases in leaf size than larger-leaved species. Similar statistical relationships were obtained from literature-derived data (F = 5.523, p = 0.030), although overall support for leaf gigantism was weaker (58%, 11/19 species pairs). A significant effect of growth form was not observed (F = 1.003, p = 0.385), and growth form did not interact with mainland stature (F = 0.441, p = 0.651), indicating size changes were similar among growth forms.

gigantism in island plants decreased with mainland leaf size. Species with smaller leaves on the mainland tended to shower stronger evidence for leaf gigantism than species with larger leaves.

Several processes may select for large leaf size in island plants. Previous research has shown that microphylly can deter mammalian herbivores (Hanley et al. 2007). Therefore, after colonising an island devoid of mammalian herbivores, plants may be released from selection for small leaf size. If microphylly is also a deterrent to toothless browsers (i.e., birds and tortoises), similar circumstances might occur on smaller islands that acquired their floras from larger oceanic islands nearby that housed toothless browsers (e.g., the Chatham Islands). If large leaves are favoured by island environments, leaf gigantism may result from relaxed selection from herbivores (Wright et al. 2017).

Results reported in Box 5.2 also indicate that insular changes in leaf area may be more complicated than simple, unidirectional selection for leaf gigantism. Increases in leaf size were related to leaf size itself. Field data indicate that larger leaves showed comparatively little change on islands, while smaller leaves showed pronounced size increases. Literature-derived data showed a similar pattern, but, in this instance, some evidence for leaf dwarfism was observed.

Seed Size

Relative to plant stature and leaf area, more is known about insular changes in seed size. Many of the studies reviewed in chapter three quantified seed sizes to determine how it might influence dispersal potential. Of the ten species listed in Table 3.1, all but two exhibited increases in seed size on islands, regardless of how seed size may influence dispersal capacity. Recall also that island bryophytes across the globe also produce larger, asexual diaspores more frequently than continental bryophytes (Patiño et al. 2013).

Burns (2016b) compared the size and shape of seeds produced by *A. ruscifolia* between Lord Howe Island and Australia and found that island plants tended to produce larger seeds. However, seed shape changed as well. Larger seeds produced by island plants were more oblong in shape than their mainland counterparts. Insular changes in seed shape could be an adaptation to facilitate bird dispersal (Lord

2004). As seed size increases on islands, a more oblong shape might make them easier for birds to swallow, which would likely increase the number of bird species that could serve as viable dispersal vectors (Burns 2013b).

Kavanagh and Burns (2014) tested for differences in seed size of 40 island–mainland taxonomic pairings from a range of islands in the South-west Pacific. Data were collected by hand wherever possible and otherwise gleaned from the literature. Results showed consistent support for insular seed gigantism, even after controlling for differences in growth form. However, seed gigantism was more prevalent in fleshy-fruited species than in those producing dry fruits, perhaps because seed size constrains the dispersal potential in dry-fruited, anemochorous species more than fleshy-fruited species.

All available evidence indicates that seed sizes increase convergently on islands. Therefore, seed gigantism appears to be a repeated pattern in island evolution. Results reported in Box 5.3 are consistent with this generalisation. However, no evidence for the island rule was observed.

A variety of processes may select for larger seeds on islands. Sea-swept costs of dispersal could select for larger seeds to reduce dispersal distances, but this seems unlikely (see Chapter 3). Alternatively, because islands house fewer species than the mainland, densities of conspecifics are often higher than mainland populations (i.e., density compensation, MacArthur et al. 1972). Seeds produced by island plants are therefore more likely to be dispersed in close proximity to a conspecific, all else being equal. Seedlings emerging from large seeds often grow faster and are more competitive than those emerging from smaller seeds (see Westoby et al. 1996). Although other processes cannot be ruled out, seedling competition, and, in particular, greater levels of symmetric competition among conspecifics can potentially explain insular seed gigantism.

BOX 5.3   Seed size in the South-west Pacific

To investigate consistent evolutionary changes in the size of seeds produced by plants inhabiting islands surrounding New Zealand, two data sets were used to test the 'island rule' (see Supplementary Resources 5.1 and 5.2). The first is comprised of field measurements of 17 pairings between insular taxa and their closest ancestor on the New Zealand 'mainland' (see Heenen et al. 2010). The second is comprised of seed size estimates obtained from the literature for 23 species inhabiting the Chatham Islands. Endemic species that originated via cladogenesis were removed to promote independence among replicates. Statistical tests of the island rule were conducted as in Box 5.1.

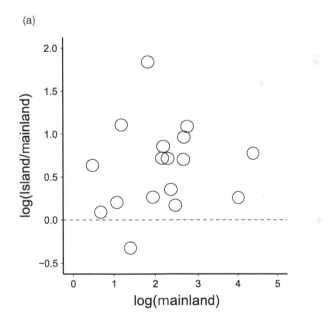

FIGURE B5.3 Evidence for gigantism in seed sizes of plants inhabiting small islands surrounding New Zealand. Island-mainland ratios of seed size (y-axis) are plotted against mainland values (x-axis) in both analyses. (a) Averages of field measurements for 17 island-mainland species pairs. (b) Seed sizes for 23 taxonomic pairings obtained from the literature (circles: fleshy-fruited species; cross-hatches: dry-fruited species).

BOX 5.3   **(cont.)**

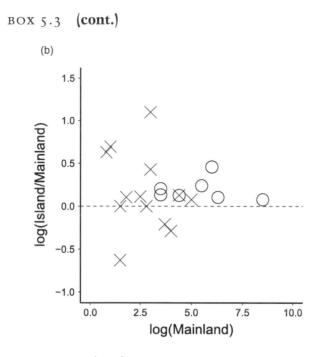

FIGURE B5.3 *(cont.)*

Results from field data showed strong support for seed gigantism and no support for the island rule (Fig. B5.3). Field data showed that 94% of taxonomic pairings (16/17) increased in size on islands, but insular increases in seed size were unrelated to seed size on the mainland (F = 0.276, p = 0.607). Similarly, literature-derived data showed that 75% of taxonomic pairings (15/20) increased in size on islands, and island–mainland seed size ratios were again unrelated to mainland seed sizes in literature-derived data (F = 0.391, p = 0.540). A significant effect of dispersal mode was not observed (F = 0.472, p = 0.502) and dispersal mode did not interact with mainland seed size (F = 0.122, p = 0.731). However, all seven fleshy-fruited species increased in seed size, while dry-fruited species exhibited both dwarfism and gigantism, suggesting stronger support for seed gigantism in bird-dispersed species.

CONCLUSIONS

Much is known about insular woodiness. Yet a definitive test of whether secondary woodiness is more likely to evolve on islands has yet to be conducted. Perhaps more importantly, does insular woodiness arise as a passive by-product of selection for increased stature, as predicted by Charles Darwin? Few studies have tested for repeated patterns in the evolution of plant traits other than woodiness, namely plant stature, leaf area, and seed size. The limited number of studies to date on the stature of island plants has found support for both dwarfism and gigantism. Results reported here suggest that changes in plant stature may follow the island rule, although much more comparative work is needed before any generalisations can be made. Seed sizes regularly increase in island organisms, providing a remarkable example of evolutionary convergence. Leaf area also exhibits a trend towards gigantism, which is weakly linked to overall leaf size. Strikingly, next to nothing is yet known about the mechanisms responsible for insular size changes in plants. Therefore, a mechanistic understanding of why plants might change in size on islands awaits further study.

6   Loss of Fire-Adapted Traits

Few terrestrial ecosystems are free from the effects of fire. Yet, until recently, wildfires were an underappreciated feature of terrestrial ecology. Perhaps as a result, fire adaptations do not feature prominently in previous discussions of how plants evolve on isolated islands. This chapter reviews a small body of emerging research on changes in fire-adapted traits in island plants, which collectively suggest that their loss on islands might be commonplace. It also indicates that future study into the relaxation of fire adaptations on islands could provide key insights into how fire might shape plant evolution on continents.

THE GEOGRAPHY OF FIRE

Wildfires occur at least occasionally in even the coldest and wettest environments on earth, but they occur regularly in savannas and Mediterranean scrublands, where they are a dominant ecological process (Chuvieco et al. 2008). The evolutionary histories of many plant species are therefore strongly intertwined with fire (Keeley et al. 2011).

Global analyses of the environmental correlates of wildfires have revealed a complex relationship between fire, temperature, and precipitation (Krawchuk et al. 2009; Archibald et al. 2018). Hot, dry ecosystems (e.g., savanna, Mediterranean scrubland) tend to burn more regularly than cold, wet ecosystems (e.g., tundra, rainforest). However, cold-temperate coniferous forests (i.e., taiga) burn regularly, and deserts, which are obviously hot and dry, are less prone to wildfires.

A key predictor of the global incidence of wildfires is net primary productivity (Krawchuk et al. 2009). Fuel (e.g., flammable leaf litter) is required for fires to burn, so, temperature and precipitation aside, more productive ecosystems tend to burn more regularly than less productive ecosystems. Seasonality is also important. Areas that experience pronounced annual variability in temperature and rainfall are more susceptible to fires. For example, savannas typically experience annual periods of increased rainfall and productivity, when fuel loads build up, which are then followed by less productive, drier periods when these fuels can ignite.

Wildfires also need an ignition source. A variety of factors can start fires, including lightning strikes, rock falls, and volcanic activity. However, in the Anthropocene, most fires are ignited by humans (Moritz et al. 2012).

As fire propagates out away from an ignition source, it can have far-reaching implications. Any point in a contiguous landscape can be connected by fire to distantly located points by a single ignition event. However, this is not the case for landscapes that are naturally fragmented by fire breaks, or landscape-level gaps in flammable material. Most notably, fire obviously cannot propagate through water, so water bodies such as lakes, rivers, and the sea form natural fire breaks.

Given the environmental determinants of wildfires, isolated islands are among the least susceptible places on earth to burning (see Wardle et al. 1997). Recall from Chapter 1 that islands tend to experience higher rainfall and lower temperatures, as well as low climatic seasonality. Islands, by definition, are also surrounded by water, so they are buffered from the effects of fire propagating out from ignition sources located far away. Given these environmental circumstances, fire is not an integral component of most island ecosystems and plant adaptations to fire may lose their relevance on islands.

FIRE-ADAPTED TRAITS

Plants have evolved a variety of traits that help them cope with wildfires (see Lawes et al. 2014). One important trait is thick bark (Pausas 2015), which can insulate live tissues from heat damage.

Another fire-adapted trait that can help plants resist the harmful effects of fire is canopy seed storage, or serotiny (Causley et al. 2016; Fig. B1.3). Species with serotinous fruits or cones enclose their seeds in thick, woody structures that insulate seeds from heat damage. They can remain closed on parent plants for long periods, opening only after they are exposed to high heat. By storing seeds in protective structures that open following wildfires, serotiny not only helps seeds resist heat damage, but also synchronises seed dispersal to coincide with periods that are beneficial for recruitment (e.g., ash-enriched soil). Seeds produced by non-serotinous species are often exposed to fire after being released from parent plants, potentially leading to their mortality. However, many plant species in fire-prone ecosystems produce seeds that are resistant to heat and smoke damage (Crosti et al. 2006; Keeley et al. 2011). Fire-adapted seeds can also lay dormant for long periods prior to exposure to heat shock or exposure to smoke (Santana et al. 2010).

In addition to becoming resistant to fire damage, plants can also evolve to be resilient to fire. Arguably the most common fire-resilient trait is the capacity to resprout (Pausas & Keeley 2014). Once damaged by fire, plants can initiate new growth in several ways. One way is to resprout from lateral meristems. Another is to resprout from special-ised, underground storage structures called *lignotubers* (James 1984; Paula et al. 2016; Lamont et al. 2017). Unlike resprouting from lateral meristems, which can be initiated by a variety of processes other than fire, lignotubers are more directly associated with fire and occur more frequently in plants that inhabit fire-prone ecosystems.

Thick bark, serotiny, and lignotubers all help plants cope with the harmful effects of fire, but it is often difficult to establish whether each has evolved specifically in response to fire (Lamont & He 2017). For example, tree bark is a multipurpose trait that has evolved to protect trees from a variety of stressors other than fire, including pathogenic attack and mechanical damage. This makes it difficult to establish whether it has evolved in direct response to fire, or some other factor (see Bradshaw et al. 2011; Keeley et al. 2011). Regardless of their precise evolutionary origin, thick bark, serotinous cones, fire-adapted seeds, and the capacity to resprout can help plants resist fire damage, or be resilient to damage by wildfires when they occur.

SYNDROME PREDICTIONS

This chapter explores a previously underappreciated pattern in island evolution – the loss of fire-adapted traits. Its predictions are similar to defence displacement. Several plant traits, including thick bark, post-fire resprouting, serotiny, and fire-adapted seeds, are known to help plants cope with wildfires on continents. Given that the climatic factors promoting wildfires are greatly diminished on most islands, island immigrants that possess fire-adapted traits should lose them evolutionarily in relaxed fire regimes on islands (Lahti et al. 2009). This chapter reviews a growing body of work testing the *loss of fire-adapted traits* hypothesis, and presents a new test for the loss of serotiny in an endemic species from Lord Howe Island.

HYPOTHESIS TESTING

Isla Guadalupe

Wildfires are an integral part of the ecology of scrublands and stunted woodlands along the west coast of North America (Naveh 1975). In contrast to the mainland, naturally occurring wildfires are rare on Isla Guadalupe, a volcanic island located 240 km west of Baja California.

Tecate cypress (*Cupressus forbesii*: Cupressaceae) is a small tree that inhabits scrublands and open forests of Southern California and Northern Mexico. At some point in the evolutionary past, an ancestor of tecate cypress dispersed to Isla Guadalupe from the mainland and slowly evolved into a new species anagenically. This newly derived species is known as Guadalupe cypress (*Cupressus guadalupensis*) and can only be found on Isla Guadalupe.

Garcillán (2010) compared fruit and seed morphology between *C. forbesii* and *C. guadalupensis* to examine whether the mainland taxon is more serotinous than the island taxon. Results showed that, in the absence of fire, over 90% of cones produced by *C. guadalupensis* opened within 200 days after maturity. Over the same period, approximately 30% of *C. forbesii* had opened and extended observations showed that only 50% had opened nearly 400 days after they matured. Garcillán (2010) concluded that there was relaxed selection for serotiny in *C. guadalupensis* in the absence of fire.

Monterey pine (*Pinus radiata*: Pinaceae) is another fire-adapted conifer that occurs both on Isla Guadalupe and along the west coast of North America. Stephens and Libby (2006) compared bark thickness of the population of *P. radiata* on Isla Guadalupe to several populations growing on the mainland. Island populations of *P. radiata* tended to have thinner bark than mainland populations, providing additional support for the loss of fire-adapted traits hypothesis from Isla Guadalupe.

California Islands

Approximately 500 km north of Isla Guadalupe lies the California Islands, which also experience weaker natural fire regimes than the nearby scrublands of Southern California (Naveh 1975, Fig. 6.1). However, unlike Isla Guadalupe, the California Islands have had a long and variable history of anthropogenic wildfires. Some of the California Islands, such as Isla Santa Cruz and Isla Santa Rosa, were burned regularly by humans after their arrival approximately 13,000 years ago.

Bishop pine (*Pinus muricata*: Pinaceae) is a fire-adapted tree species that occurs along the west coast of North America, in addition to several California Islands, including Isla Santa Cruz and Isla Santa Rosa. Stephens and Libby (2006) compared bark thicknesses of *P. muricata* populations on the islands to populations on the mainland. They found that bark thicknesses in *P. muricata* did not differ between islands and the mainland. Therefore, their results were inconsistent with the loss of fire-adapted traits hypothesis. However, they suggest that a clear test of the loss of fire-adapted traits hypothesis is confounded by anthropogenic burning and that *P. muricata*

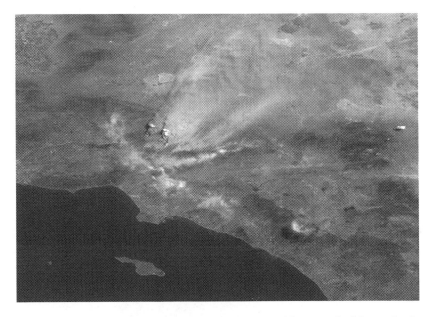

FIGURE 6.1 Satellite image of a wildfire in California and offshore islands that were buffered from the effects of fire by the ocean. (photo from the Universal History Archive/Getty Images)

could have adapted to anthropogenic burning since the arrival of people at the end of the last glacial maximum.

The germination behaviour of seeds produced by many shrub species inhabiting the scrublands of coastal California is influenced by fire. In a thorough study of many shrub species, Keeley (1987) found charate-stimulated seed germination in around a quarter of the local species pool. Heat-stimulated germination was observed in another quarter of these species. Therefore, approximately half of the species in the coastal scrublands of California produce fire-adapted seeds. The remainder germinated vigorously in the absence of fire-related cues, however, indicating that a significant fraction of the flora is adapted to recruit following other forms of disturbance.

Carroll et al. (1993) tested whether seeds produced by island populations of several shrub species were more poorly adapted to fire on Isla Santa Cruz than on the mainland. They collected 400 seeds

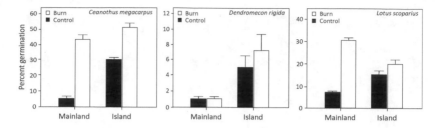

FIGURE 6.2 Results from seed germination trials in three shrub species that inhabit the California Islands and the surrounding mainland (data from Carroll et al. 1993). Mean percent germination (y-axis) of seeds collected either from Isla Santa Cruz or the adjacent mainland (x-axis) are shown (±SE, n = 400 seeds). White bars represent seeds that were subject to fire prior to sowing. Black bars represent controls.

from three species (*Ceanothus megacarpus*: Rhamnaceae; *Dendromecon rigida*: Papaveraceae and *Lotus scoparius*: Fabaceae) on both the California Islands and the adjacent mainland. Seeds were sowed in eight replicate pots, half of which were then then covered by 4 cm of a flammable, fire-adapted chaparral species (*Adenostoma fasciculatum*: Rosaceae), which was then ignited and allowed to burn thoroughly. The other containers were left unburned as controls.

Results showed that burning enhanced germination in general, both in island- and mainland-sourced plants (Fig. 6.2). However, in two out of the three species (*C. megacarpus* and *L. scoparius*), the difference in germination rates between treatments was greater in mainland- than island-sourced plants. Consistent with the loss of fire-adapted traits hypothesis, the control (unburned) seeds of island plants germinated more readily than mainland plants.

Québec Islands

Wildfires are a common phenomenon in northern coniferous forests (i.e., taiga, Rogers et al. 2015). Another conspicuous feature of boreal ecosystems is the widespread occurrence of lakes. As glaciers retreated during Pleistocene interglacial periods, they left depressions in the landscape that have since filled with water. This process has

FIGURE 6.3 A serotinous jack pine cone (*Pinus banksiana*: Pinaceae). photo taken by Ed Reschke/Getty Images

resulted in thousands of lakes, many of which contain islands that support conifer populations.

Detailed analyses of past fire events in these landlocked, boreal islands show that they experience different fire regimes than the surrounding mainland. Fires are less frequent on the mainland, but, when they do occur, they are very destructive. On the other hand, islands experience less lethal wildfires at more regular intervals (Bergeron & Brisson 1990).

Gauthier et al. (1996) and Briand et al. (2015) investigated the degree of serotiny in populations of jack pine (*Pinus banksiana*; Fig. 6.3) on islands in Lake Duparquet, Québec, Canada. They tested the hypothesis that island populations of jack pine produce less serotinous cones than nearby populations on the mainland. Their results were generally consistent with the loss of fire-adapted traits hypothesis – serotiny was more prevalent in mainland populations. They conclude that lethal fires on the mainland have selected for serotiny to promote reseeding following mass mortality events. On the other hand, more continuous seed release is favoured in island populations that experience lethal fires less frequently.

The timescale involved in the loss of serotiny in jack pine may have important implications for our understanding of how human-induced fires may influence the evolutionary trajectories of island plants. Islands in Québec are of recent geologic origin. The region was covered in ice during the last glacial maximum, so most boreal islands originated relatively recently, following glacial retreat. Therefore, changes in fire-adapted traits appear to have occurred relatively rapidly, within the last 10,000 years. This timeframe is consistent with the expansion of humans across much of the planet, which suggests that anthropogenic activity could potentially select for the loss of fire-adapted traits elsewhere.

Canary Islands

Mountainous regions of the western Canary Islands (Islas Gran Canaria, Tenerife, Hierro, and La Palma) support conifer forests that are dominated by a single endemic species, *Pinus canariensis*: Pinaceae (Fig. 6.4). Unlike most other oceanic islands, wildfires regularly occur on the Canary Islands (see Climent et al. 2004). Much of the archipelago experiences hot, dry conditions seasonally, and volcanism is commonplace, providing a naturally occurring ignition source for wildfires. Anthropogenic fires are also commonplace.

Climent et al. (2004) quantified bark thickness and levels of serotiny across multiple populations of *P. canariensis* growing on

FIGURE 6.4 Canary Island pine (*Pinus canariensis*: Pinaceae) on La Palma Island.

several islands. They found that population-level variation in bark thickness and serotiny were related to fire frequencies. Bark thickness averaged approximately 35 mm (22–49 mm), and roughly 19% (3–35%) of all trees produced serotinous cones. Both variables increased with fire frequencies as well as several environmental correlates of fire.

The phylogenetic history of *P. canariensis* is somewhat ambiguous. Some phylogenetic analyses indicate that stone pine (*Pinus pinea*) is its closest ancestor. *P. pinea* is native to Southern Europe and the Mediterranean basin (Wang et al. 1999), and, in sharp contrast to *P. canariensis*, it shows no evidence for canopy seed storage (Tapias et al. 2001). Its average bark thickness in south-eastern France averages 22 mm (9–44 mm), which is substantially lower than *P. canariensis* (Rigolot 2004). Under this phylogenetic hypothesis, *P. canariensis* has gained, rather than lost, fire-adapted traits.

Other phylogenetic hypotheses indicate that *P. canariensis* is most closely related to maritime pine (*Pinus pinaster*), which commonly occurs in coastal regions of the Iberian Peninsula and the western Mediterranean basin (Liston et al. 1999). *P. pinaster* produces highly serotinous cones that persist on parent plants for up to 40 years (Tapias et al. 2004), whereas in *P. canariensis*, cones remain closed on parent plants for no more than 10 years (Tapias et al. 2004). So determining whether *P. canariensis* has gained or lost fire-adapted traits since colonising the Canary Islands hinges entirely on the underlying phylogenetic hypothesis. Nevertheless, it is clear that population-level variation in fire-adapted traits in *P. canariensis* is linked to variation in fire frequencies within the Canary island archipelago.

New Zealand

Wildfires were uncommon in New Zealand prior to human arrival (Perry et al. 2014). In stark contrast, 2,000 km west of New Zealand lies one of the most fire-prone places on the planet, Australia (Bradstock et al. 2012; Clarke et al. 2015). Most of New Zealand's flora is derived from over-water dispersal from Australia (see Gibbs 2006), providing an ideal situation to test for the loss of fire-adapted traits. By and large, the New Zealand flora is poorly adapted to frequent wildfires (Perry et al. 2014). However, support for the loss of fire-adapted traits hypothesis is not straightforward.

Lawes et al. (2014) measured bark thickness in a total of 1,246 temperate rainforest trees belonging to 51 species in both Australia and New Zealand and used these resulting data to test whether bark thickness was greater in Australia than in New Zealand. Results were equivocal. While New Zealand trees have thinner bark on a global scale, they tended to have thicker bark than their Australian counterparts. However, Lawes et al. (2014) are quick to point out that bark thickness is just one of many plant traits that can help plants mediate the harmful effects of fire. For example, the capacity to resprout after fire is a common mechanism that promotes plant resilience to wildfires (Bond & Midgley 2001). While post-fire resprouting is a common feature of Australian trees (Knox & Clark 2012; Clarke et al. 2015), very few New Zealand tree species resprout successfully after wildfires (Wiser et al. 1997).

Manuka (*Leptospermum scoparium*: Myrtaceae) is arguably New Zealand's most fire-adapted species. It is also New Zealand's most widespread woody plant and is particularly common in areas recently burned by humans, as well as naturally occurring wetlands (Stephens et al. 2005). It also occurs across a wide range of habitats in Australia, particularly in places that are exposed to frequent wildfires, as well as wetlands. Australian populations often exhibit marked serotiny, while serotiny is generally less common in New Zealand (Bond et al. 2004b).

Examination of the seed germination behaviour of *L. scoparium* showed that germinability was not related to flammability or serotiny (Battersby et al. 2017a). On the other hand, *L. scoparium* produces lignotubers in Australia, but not New Zealand (Bond et al. 2004b). Similar to *P. canariensis*, levels of serotiny also vary widely across New Zealand (Harris 2002), and generally increase with temperature and decline with latitude and rainfall (Battersby et al. 2017b). So while *L. scoparium* appears to be more fire adapted in Australia, New Zealand populations do possess fire-adapted traits, which vary among regions according to their susceptibility to wildfires.

Lord Howe Island

The subtropical rainforests of Lord Howe Island are filled with plant species that dispersed over-water from Australia. Given the incidence of fire in Australian ecosystems, it is not surprising that several species endemic to Lord Howe Island (*Leptospermum polygalifolium*: Myrtaceae and *Melaleuca howeana*: Myrtaceae, Fig. 6.5) share a recent common ancestor in Australia that produces serotinous fruits.

To test whether canopy seed storage is reduced on Lord Howe Island relative to Australia, levels of serotiny were compared between *M. howeana* and a close mainland relative, *Melaleuca ericifolia* (see Edwards et al. 2010). Results showed clear evidence for the loss of serotiny in *M. howeana* (Box 6.1). Over 90% of second-year fruits were serotinous on the mainland, while less than 10% were serotinous on the island.

FIGURE 6.5 Non-serotinous fruits of *Melaleuca howeana*: Myrtaceae on Lord Howe Island.

BOX 6.1   **Loss of serotiny in *Melaleuca***

Melaleuca ericifolia: Myrtaceae, or swamp paperbark, is a common tree in Eastern Australia that produces serotinous fruits that contain numerous small seeds. It is the closest living relative of *Melaleuca howeana*, a similar species that is endemic to Lord Howe Island (see Edwards et al. 2010). At some point in the evolutionary past, an ancestral population of *M. ericifolia* dispersed to Lord Howe Island and subsequently diverged into a separate species.

Eastern Australia and Lord Howe Island have very different fire regimes. While fire is an integral component of the ecology of southeast Australia, it is virtually absent from Lord Howe Island. Therefore, in the absence of fire, selection may have favoured the loss of serotiny in the insular taxon (Lahti et al. 2009). To test for the loss of serotiny in *M. howeana*, the fraction of serotinous fruits on a single branch from 20 individuals of *M. ericifolia* in Abrahams Bosom Reserve in coastal New South Wales (–35.01613, 150.836004) was quantified (n = 136 fruits) and compared to an identical fruit count on 20 individuals of *M. howeana* on Lord Howe Island (n = 151 fruits).

Results showed that most fruits produced by *M. ericifolia* (98.5%) were serotinous (Fig. B6.1). On the other hand, far fewer serotinous fruits were produced by *M. howeana* (12.6%, X^2 = 305, p < 0.001). Results are therefore consistent with the hypothesis that selection favours the loss of fire-adapted traits on fire-free islands.

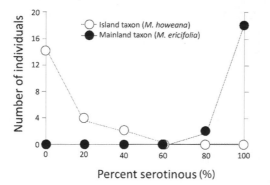

FIGURE B6.1 Differences in serotiny between *Melaleuca howeana* (open circles, dashed line), which is endemic to Lord Howe Island, and *Melaleuca ericifolia* (closed circles, solid line), a close relative on the Australian mainland.

Although this result clearly supports the loss of fire-adapted traits hypothesis, ecological processes other than fire may select for serotiny on the mainland. *M. ericifolia* is commonly known as swamp paperbark and frequently occurs in wetlands. Hamilton-Brown et al. (2009) investigated the fruiting behaviour of *M. ericifolia* in Australian wetlands and found that in the absence of fire seed release is timed to coincide with seasonal periods of environmental conditions that are more conducive to seedling recruitment. Given that wetlands are uncommon on Lord Howe Island at present, the loss of serotiny might be associated with differences in habitat between the island and the mainland, rather than fire.

CONCLUSIONS

Although relatively few studies have tested the loss of fire-adapted traits hypothesis, previous work has often found support for its predictions. Conifers on Isla Guadalupe exhibit reduced bark thickness and serotiny relative to the mainland populations in North America. Several shrub species on the California Islands show reduced seed survivorship and germination capacity when exposed to fire. Serotiny is reduced in conifer populations on islands in glacial lakes in the Canadian taiga, and paperbark populations on Lord Howe Island show reduced levels of serotiny relative to their mainland relatives in Australia.

Other islands exhibit more complex patterns in fire-adapted traits. Comparisons of bark thickness between the California Islands and the mainland failed to uncover substantial differences. Contrary to syndrome predictions, bark thicknesses in New Zealand trees were not dramatically different from those of Australian trees. Although New Zealand populations of *L. scoparium* appear to lack lignotubers, and seem to be less serotinous, formal quantitative comparisons between New Zealand populations and their Australian ancestors have yet to be conducted. Difficulties in establishing the phylogenetic history of the *P. canariensis* prohibits firm conclusions regarding whether it has gained or lost fire-adapted traits in Macaronesia.

Within archipelago studies of spatial variation in fire-adapted traits in New Zealand and the Canary Islands found clear associations between local fire frequencies and population-level variation in fire-adapted traits. Serotiny in *L. scoparium* varies across New Zealand according to environmental correlates of fire and local fire histories. Similarly detailed study among Australian populations of *L. scoparium* would provide an insightful comparison of the interplay between local and regional correspondence of fire regimes and fire-adapted traits. Provided an unambiguous understanding of the phylogenetic history of *P. canariensis* can be obtained, similar comparisons of regional variation in fire-adapted traits between the Canary Islands and the mainland would be equally informative.

Overall results suggest that the loss of fire-adapted traits could very well be a component of the island syndrome in plants. However, previous work is comprised of mostly single-species studies that come from just seven islands, which advocates a cautious approach towards general conclusions. Additional work on a greater number of species from other archipelagos is needed before firm generalisations can be reached.

Several confounding effects provide a serious challenge for future work. First, multiple traits can be deployed simultaneously to provide either resistance or resilience to fire. So investigations involving just one trait, which characteristics virtually all previous work, does not rule out the existence of opposing patterns in others. Future work would benefit from investigating multiple fire-adapted traits simultaneously. Not only would this generate a more wholistic understanding of fire-related selection pressures, but it would also generate a better understanding of which traits are more likely to diverge on less fire-prone islands. Another serious challenge is the potential effects of anthropogenic fire histories on the expression of fire-adapted traits. Given that many archipelagos were discovered only recently, and have a shorter history of anthropogenic burning, plants with longer generation times might not have had enough time to adapt to increased fire frequencies. However, if these issues can be overcome, future work on the loss of fire-adapted traits might not only inform our understanding of repeated patterns in plant evolution on islands, but they might also provide key insights into our understanding of how plants adapt to fire in general.

7   **Conclusion**

Emblematic Island Plants

Visiting the Natural History Museum in London is a fantastic experience. An enormous diversity of objects is on display, from giant dinosaur fossils, to butterflies collected by Alfred Russel Wallace, and the famous marble statue of the man who made sense of it all, Charles Darwin. However, one particular type of animal is displayed repeatedly, in far greater frequency than their occurrence in the wild. Island 'oddities' take centre stage.

A stuffed kiwi bird sits prominently among the skins of other bird species collected from across the globe. The piece of subfossilised leg bone that Sir Richard Owen correctly identified as belonging to a giant flightless bird from New Zealand has its own glass cabinet, sitting alongside a marble statue of Sir Richard Owen himself. In addition to two stuffed specimens, a fully articulated dodo skeleton is on display. In fact, the distributional range of the dodo in the museum is not restricted to the collection space. It extends into the gift shop, where its comical appears adorns plates, cups, tea towels, and even gift cards.

Island oddities clearly captivate the human imagination. By breaking the rules governing 'normal' types of organisms on continents, they exemplify the full breadth of forms that evolution can produce. However, only island animals are displayed. Island plants are nowhere to be seen. Or, at least if they are displayed, they don't take centre stage, and I haven't been able to find them over the course of several visits. This is a shame in my opinion, as many island plants depart radically from their mainland counterparts, and perhaps even surpass the dodo in their odd appearance and capacity to captivate.

The following descriptions of five emblematic island plant species typify the island syndrome in plants. Selecting just five was not easy. But when these five species are considered collectively, they exemplify the phenomena explored in previous chapters.

CUCUMBER TREE (*DENDROSICYOS SOCOTRANUS*: CUCURBITACEAE)

Although few of us have encountered this type of cucumber in the wild, we regularly encounter its family members when we go to the market. Watermelons, squash, and pumpkins all belong to the Cucurbitaceae, the vast majority of the which are herbaceous vines. The ground-dwelling growth habit of this family has facilitated its most distinctive aspect, the evolution of large, heavy fruits. Because their fruits don't require structural support to hold them aloft, edible species have been bred selectively by humans to be more than a meal.

Rather than being restricted to vines and lianas, heavy fruits and seeds were once thought to have been an ancestral condition in all angiosperm trees. The *Durian Theory* argues that in order to support big fruits, primitive trees evolved a distinctive, *pachycaul* growth form (Corner 1949). The term 'pachycaul' refers to trees with oversized parts – large fruits and seeds that are attached to thick stems and ultimately massive, unbranched trunks. They also have relatively large leaves, which are often deeply lobed or compound to dissipate shear stress.

We now know that the pachycaul growth form is not the ancestral state of modern angiosperms, mainly because there is little evidence for this hypothesis in the fossil record. However, pachycaul trees are a conspicuous component of many island floras. They are also common in island-like habitats such as the summits of tropical mountains (Mabberley 1979). However, arguably the most impressive example of the pachycaul growth form is the cucumber tree (*Dendrosicyos socotranus*: Cucurbitaceae) of Socotra, the only arborescent member of the gourd family. It has an awkward, dumpy appearance, with a massive trunk, weeping (pendant) branches and large palmate leaves (Fig. 7.1). It epitomises the pachycaul growth form and may be the best example of insular woodiness on the planet (see Olson 2003).

FIGURE 7.1 A cucumber tree
(*Dendrosicyos socotranus*:
Cucurbitaceae) on Socotra Island.
(photo from Getty Images)

Molecular phylogenetic work indicates that it diverged from its closest living ancestor 14–30 Ma (Schaefer et al. 2009). Two decades ago, geologists estimated that Socotra was around 10 million years old (Ghebreab 1998). It therefore appeared that *D. socotranus* evolved its unusual appearance when Socotra was still part of Africa, and rather than being an example of insular woodiness, its stature became 'relictual' after it subsequently went extinct in mainland Africa. However, more recent geological work casts doubt over whether its pachycaul growth form is in fact an ancestral trait.

Culek (2013) reinterpreted the geogologic history of the region and concluded that the Gulf of Aden opened 17–20 Ma, much earlier than previously thought. Under this sencerio, the geologic origin of Socotra is roughly coincident with the divergence time between *D. socotranus* and its closest mainland relatives. Therefore, the unusual, pachycaul growth form of *D. socotranus* may instead be the result of insular selection pressures.

Socotra's cucumber tree (*D. socotranus*) is a distinctive example of island gigantism and possibly insular woodiness. Its large, stumpy, almost comical appearance compared to its diminutive, mainland ancestors is rivalled only by the dodo and other flightless birds. It is also a poignant reminder that our understanding of island evolution often hinges on accurate phylogenetic and geologic hypotheses, which can be uncertain and subject to change.

ELEPHANT CACTI (*PACHYCEREUS PRINGLEI*: CACTACEAE)

Cacti are among the most distinctive plants on the planet. They also come in all shapes and sizes. Some cacti are epiphytes and grow at the tops of trees (e.g., *Rhipsalis*), while others just barely emerge from the ground and look like pebbles (e.g. *Lophophora*). At the other end of the size spectrum is *P. pringlei*, commonly known as giant elephant cacti, or cardón in Spanish. Not only can it grow to nearly 20 m tall, its flowers are also strikingly large and distinctively bell shaped.

P. pringlei is a common component of island plant communities in the Sea of Cortez, a slender finger of water that separates Baja California from mainland Mexico. Most islands in the Sea of Cortez were connected to the mainland during the last glacial maximum, and were subsequently separated from the mainland by rising sea levels. However, a smaller number of islands are oceanic in origin and have never been connected to the mainland. There is a long history of research into the island biogeography of the archipelago and much is known about their plant communities (Cody 2002; Felger & Wilder 2012).

Populations of *P. pringlei* on islands in the Sea of Cortez differ from those on the mainland in several emblematic ways. Arguably the most striking difference between island and mainland populations is their stature. On highly isolated islands, such as Isla Santa Cruz and Isla San Pedro Martir, *P. pringlei* is markedly dwarfed (Cody 1984, 2002; Wilder & Felger 2012). Unlike populations on the mainland, which are typically rooted to the ground by a single stem, island populations typically lack a trunk and begin to branch much earlier in ontogeny (Figs 7.2 and 7.3). However, the extent to which its dwarfed stature is

(a)

(b)

FIGURE 7.2 (a) A mainland population of *Pachycereus pringlei*: Cactaceae on Baja California Sur, Mexico. (photo from Getty Images) (b) A dwarfed population on Isla Santa Catalina in the Sea of Cortez. (photo from Getty Images)

genetically determined, and an explanation for why insular dwarfism might be selectively advantageous, awaits additional study.

Another distinctive attribute of island populations of *P. pringlei* is their reproductive biology. *P. pringlei* has a remarkably labile breeding system, with populations being comprised of individuals that can be male, female, or hermaphrodite. Strikingly divergent patterns in the geographical distribution of self-compatible hermaphrodites occur on either side of the Sea of Cortez (Fleming et al. 1998). On the Mexican mainland, the proportion of hermaphroditic plants declines with latitude ($t = -2.427$, $p = 0.041$). While in populations on the Baja Peninsula, hermaphroditism increases with latitude ($t = 3.850$, $p = 0.002$).

Regardless of which mainland population they are compared to, island populations have much higher frequencies of hermaphrodites (Fig. 7.3), as predicted by Baker's law. Although much work has been done on the breeding system of *P. pringlei* (Fleming et al. 1994, 1998; Molina-Freaner et al. 2003; Gutiérrez-Flores et al. 2016), it is not yet

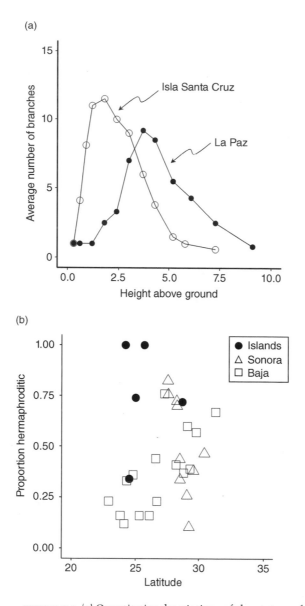

FIGURE 7.3 (a) Quantitative descriptions of the stature of *Pachycereus pringlei*: Cactaceae on Isla Santa Cruz (white circles) and the mainland (La Paz, Baja California; black circles). Mainland populations are taller and begin branching later in ontogeny than dwarfed insular populations. Data from Cody (1984). (b) Complex biogeographic patterns in hermaphroditism in *P. pringlei* across Northern Mexico, increasing with latitude in Baja California (squares) and decreasing with latitude in Sonora and Sinaloa (triangles). However, hermaphroditism is highest in insular populations in the Sea of Cortez (circles). Data from Molina-Freaner et al. (2003) and Gutiérrez-Flores et al. (2016)

clear why it varies so differently between mainland populations on either side of the Sea of Cortez. Equally unclear is whether the higher incidence of hermaphroditism on islands results from either immigrant selection or *in situ* natural selection.

Elephant cacti (*P. pringlei*) is an emblematic island plant species for several reasons. First, much like its animal namesake, *P. pringlei* provides an excellent example of insular dwarfism, every bit as striking as the dwarfed Stegodons on Flores Island. It also obeys Baker's law, which now appears to be a key component of an island syndrome in plants.

CAFÉ MARRON (*RAMOSMANIA RODRIGUESI*: RUBIACEAE)

Similar to the takahe, *R. rodriguesi* (café marron, or wild coffee in English) was once thought to be extinct. However, in this case its saviour was a schoolboy named Hedley Manan, who found a single surviving tree clinging to existence on a remote part of Rodrigues Island in the Indian Ocean. *Ramosmania* is a genus with just two species, café marron and *Ramosmania heterophylla* (Verdcourt 1996). Both are small trees restricted to the Mascarene Islands. However, *R. heterophylla* wasn't as lucky as café marron. It went extinct before it could be found by a saviour naturalist.

After the last individual of café marron was found, its fate was hardly secure. Folklore on Rodrigues has it that a brew made of its leaves can cure hangovers, and even treat venereal diseases (Magdalena 2015). So it was very nearly picked apart by desperate locals. More alarmingly, it also failed to set seed, despite decades of monitoring.

Given its precarious status, cuttings were carefully collected and flown to the micropropagation unit of Kew Gardens, where, after a great deal of effort, horticulturalists managed to establish a small *ex situ* population of genetically identical clones. These plants flowered profusely (Fig. 7.4), but they stubbornly refused to set seed. Its flowers appeared to be hermaphroditic, like many other island plants. Yet, every attempt to foster self-pollination failed. For two decades, garden

FIGURE 7.4 Café marron (*Ramosmania rodriguesi*: Rubiaceae) growing in Kew Gardens, London.

botanists tried and tried, to no avail, and it seemed that café marron was among the 'living dead', doomed to a reliance on people to perpetuate its existence.

Closer inspection of its flowers showed that they contained both viable pollen and fertile ovules. But their stigmas were attached to the ovaries by curiously short styles. Further analyses showed that the stigmas were non-receptive, which appeared to render the species terminally male (Owens et al. 1993). Given that café marron is distantly related to other branches of its evolutionary lineage, and deeply endemic to the Mascarene Islands, perhaps this is unsurprising. Baker predicted that after immigrant selection for self-compatibility, natural selection on islands should favour the evolution of outcrossing mechanisms and *R. rodriguesi* appeared to be functionally dioecious, or perhaps heterostylous.

Then something miraculous happened. This apparently hapless plant got lucky. After a prolonged a heat wave in London, its new-found home, plants began to produce flowers with longer styles, which proved to be more receptive. Viable seeds were finally produced by flowers developing under warmer conditions, and these seeds subsequently germinated (Magdalena 2015).

Three years after they germinated, some of the new plants began to produce flowers. Amazingly, some of these flowers were distinctly different from those produced by the last plant in the wild (and the army of clones created from its cuttings). This new type of flower had smaller petals, larger stigmas, and no pollen, rendering them effectively female. Cross-fertilising these flowers with pollen collected from the original flower type led to more vigorous seed set.

Subsequent observations of the seedlings that emerged from these seeds showed that café marron is dimorphic in more than just its flowers. It was as though the seeds had been mislabelled, as an altogether different looking plant came up from where they were carefully planted by horticulturalists. Instead of the large, round, green leaflets produced by adults, these plants produced distinctively narrower leaves with a mottled-brown hue and a reddish midvein. Much to the planters' relief, once they grew to approximately 1.5 m tall, they suddenly changed their appearance and began to look like their parents and produced the more familiar, adult leaf type. Rodrigues was once home to giant, herbivorous tortoises, and the juvenile leaves may have provided a degree of protection against their attack (see Eskildsen et al. 2004).

Café marron (*R. rodriguesi*) is an elegant example of several aspects of the island syndrome in plants. First, although it appears to be hermaphroditic, *R. rodriguesi* is for the most part incapable of uniparental reproduction, as Baker predicted, so that established island species avoid inbreeding depression. However, gender in this particular species is labile, and appears to be somewhat temperature dependent. Without it, café marron's fate would be sealed as one of a growing number of 'living dead' species, which without constant human care would be doomed to extinction (Magdalena 2015). It is also a stunning example of leaf heteroblasty, but it remains unclear whether this has a physiological explanation or whether the strange shape and reflectance properties of juvenile leaves are still signalling to the ghosts of herbivory past.

LANCEWOOD (*PSEUDOPANAX CRASSIFOLIOUS*: ARALIACEAE)

As Captain James Cook circumnavigated the globe, he took with him one of the best botanists that ever lived, Sir Joseph Banks. As the ship's naturalist, it was Banks' job to catalogue all the new plants they encountered along their way. However, he was totally unprepared for a small tree species that he encountered as he stepped ashore in New Zealand, despite all of his experience.

P. crassifolius (lancewood) undergoes a striking series of metamorphoses during ontogeny. Its developmental changes are so extreme that Banks initially named adults and juveniles as separate species (Allan 1982). Nothing in his experience had prepared him for the sudden changes in morphology it displays through ontogeny, changes that are mirrored by butterflies as they pupate from larvae into adults.

P. crassifolius goes through not one, but two, radical transformations as it grows up. Seedling leaves are small, soft, and narrow, and often have one or more large, lateral lobes, giving them an irregular, asymmetric shape. However, the most distinctive attribute of seedlings is their colour. Instead of appearing green to human eyes, or, in other words, having reflective properties that are consistent with the absorptive properties of chlorophyll a and b, the upper surface of their leaves is filled with other types of pigments that make them appear mottled brown. So, when viewed from above, they are hard to notice against a background of leaf litter. Quantitative analyses of their reflective properties indicate that they would have been nearly invisible to herbivorous birds (Fadzly et al. 2009).

As seedlings grow taller (~50–100 mm tall), they undergo their first morphological transition, and suddenly begin to produce leaves that are strikingly long, narrow, and incredibly rigid (Fig. 7.5). They also have pronounced lateral leaf spines along their margins, each flanked by a patch of different-coloured tissue, which increases their conspicuousness to the avian eye (i.e., aposematism; Fadzly et al. 2009). Given their strikingly fearsome appearance, it is easy to imagine how difficult they would be to swallow whole.

(a)

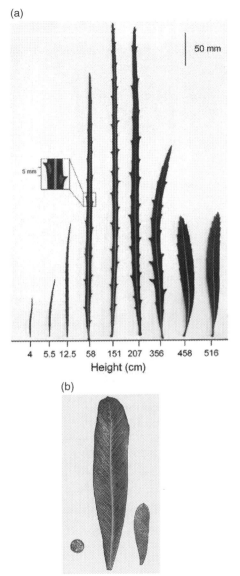

Height (cm)

(b)

FIGURE 7.5 (a) Ontogenetic shifts in the morphology of lancewood (*Pseudopanax crossifolius*: Araliaceae) leaves, with the youngest leaf positioned at left and progressively older leaves to the right. Reprinted with permission from Kavanagh et al. (2016) (b) A juvenile (left) and adult (right) leaf of *Pseudopanax chathamicus*. The coin measures 20 mm in diameter.

As plants grow taller, they undergo a second abrupt change in morphology (Fig. 7.6). Similar to most divaricately branched heteroblastic plants in New Zealand, once *P. crassifolius* grows to approximately 3 m tall, they become altogether unremarkable – adult leaves are oblong, green, and soft. They also lack lateral leaf spines and

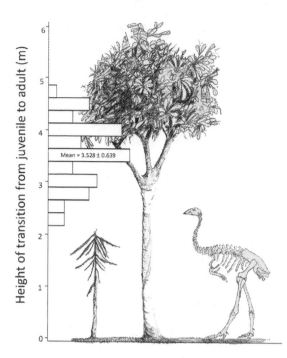

FIG 7.6 Maturation heights in lancewood (*Pseudopanax crassifolius*: Araliaceae). At sexual maturity, plants switch from producing spinescent, sapling leaves (illustrated to the left) to non-spinescent adult leaves (illustrated to the right), and plants branch for the first time. The frequency distribution represents the height of the first branching event in 50 mature trees from Nelson Lakes National Park, New Zealand. The average height at maturity (~3.5 m tall) is roughly coincident with the height of the tallest known moa species (illustrated to the right).

associated pigmentation. This transition height between saplings and adults is much taller than *R. rodriguesi* (café marron), and is consistent with the height of the tallest known browsing bird from New Zealand, rather than the vertical reach of a giant tortoise.

P. crassifolius epitomises insular leaf heteroblasty. Just like many types of butterflies, it begins life cryptically coloured and switches to become aposematically coloured once it becomes defended (i.e., spinescent). It then undergoes a final transition at maturity that is unparalleled in butterflies. Once it reaches a height refuge from herbivory, it produces leaves that appear to be structurally undefended.

Recent work illustrates that developmental shifts in colouration are not restricted to the upper surfaces of *P. crassifolius* leaves (Kavanagh et al. 2016). As plants grow taller, flightless birds would begin to look upwards at lancewood leaves, rather than downwards. So, aposematic defensive colouration on the upper side of leaves would have become increasing useless as plants grow taller, while colour-based defence on the lower surface of leaves would become increasingly useful. Concomitant with changes in the angle of detection by ground-dwelling herbivores, *P. crassifolius* undergoes a series of developmental shifts in the colouration of the lower (abaxial) surface of leaves. Seedling leaves that are only a few centimetres from the ground appear green to human eyes, but become increasingly red as they grow taller and eventually switch back to appearing green at maturity.

In the recent evolutionary past, an ancestor of *P. crassifolius* dispersed to the Chatham Islands, which, unlike New Zealand, never housed toothless browsers. There it diverged into a new species anagenically – *Pseudopanax chathamicus*. Chatham Island lancewood is nearly identical to *P. crassifolius* genetically (Heenan et al. 2010), but they are very different morphologically (Fig. 7.5). Seedlings lack pigmentation that make them cryptic against a background of leaf litter, and, although saplings produce longer leaves than adults, they are markedly less rigid, much wider, and less heteroblastic. They lack lateral leaf spines and any associated aposematic colouration.

New Zealand lancewood (*P. crassifolius*), and its close relative on the Chatham Islands (*P. chatahmicus*) exemplify a key component of the island syndrome in plants: defence displacement. It also exemplifies several putative defensive adaptations to toothless browsers. While leaf heteroblasty and leaf spinescence may be prevalent on several other toothless browser islands, additional study is needed to confirm their status as a component of the island syndrome across the globe.

COCO DE MER (*LODOICEA MALDIVICA*: ARECACEAE)

It should come as no surprise that the world's largest seed is produced by an island species. *L. maldivica* (coco de mer) is a massive palm

FIGURE 7.7 Coco de Mer (*Lodoicea maldivica:* Arecaceae) from the Seychelles, the heaviest fruit on earth. (photo from narvikk/Getty Images)

species that is endemic to the Seychelles Islands. Its most distinctive feature is an enormous fruit, a coconut that weighs up to 45 kg. European explorers found them washed ashore on beaches across the Indian Ocean, from South Africa to India. But for decades their source of origin was unknown.

Rather than the typical oval shape of water-dispersed coconuts, the coco der mer has two distinctive lobes (Fig. 7.7). Given their feminine shape, they were highly sought after as an aphrodisiac in the sixteenth, seventeenth, and eighteenth centuries (Blackmore et al. 2012). Fruits that washed ashore across the Indian Ocean sold for lavish prices all the way from China to Europe. Because their source of origin was unknown, a myth originated that they grew at the bottom of the sea, which ultimately gave rise to its name, 'coco de mer'. The mystery was finally solved by French sailors in 1769, when they stumbled into a grove on Praslin Island in the Seychelles. *L. maldivica* was not endemic to the bottom of the sea after all.

Interestingly, viable fruits of the coco de mer do not float (Gollner 1999). Only the outer casing of fruits, without viable seeds, were found washed ashore in far-flung places. In fact, viable coco de mer fruits have no apparent dispersal mechanism, aside from rolling downhill away from parents rooted on steeper slopes (Edwards et al. 2002). This suggests that the loss of dispersal capacity (i.e., smaller air bladders) could have evolved to help them remain on their island of origin in the Seychelles. On the other hand, selection may have also favoured larger seeds, without concomitant increases in dispersal aides as described by Carlquist (1974).

Insular gigantism in *L. maldivica* extends beyond the size of its seeds. It produces absolutely massive leaves. Although they vary somewhat among populations, they can grow to a whopping 10 m long and 4.5 m wide (Fleischer-Dogley et al. 2011). Leaf morphology also varies dramatically through ontogeny, for reasons that appear to be completely unrelated to herbivory. Young plants produce leaves on increasingly long petioles, stretching to 10 m long, but as plants grow taller and their trunk begins to emerge from the ground, petiole size declines (Edwards et al. 2002). Interestingly, the shape of the leaves and associated petioles trap water and nutrients and then funnels them towards the base of the plant where they can be better absorbed by the plant's root system (Edwards et al. 2015).

Coco de mer (*L. maldivica*) is arguably the finest example of insular gigantism. It produces the largest fruits of any plant on earth, which sink when immersed in water, unlike most other coconuts. It also produces massive leaves and displays marked heteroblasty, which appears to be linked to water and nutrient acquisition. So, in addition to exemplifying leaf and seed gigantism, it is also a great example of how factors other the herbivory might select for leaf heteroblasty in island plants.

SUMMARY

Differences in Defence

Island plants are often defended differently than their continental counterparts. The loss of defence occurs repeatedly in well-defended plants that colonise islands devoid of vertebrate herbivores. Island plants synthesise defensive chemicals in smaller quantities, and lose spinescent structures such as prickles, thorns, and spines. Therefore, defence displacement, or the loss of defence on herbivore-free islands, appears to be a component of an island syndrome in plants.

Plant protection mutualisms are common on continents. However, plant protectors such as ants often fail to colonise isolated islands. On islands devoid of plant protectors, plants often stop producing domatia, Müllerian bodies, and extrafloral nectaries. Although consistently observed where it has been investigated, the dissolution of ant–plant mutualisms has only been documented on a handful of occasions, precluding its status as a component of an island syndrome in plants at present.

Tests for convergent evolution of anti-herbivore traits among islands that supported avian or reptilian herbivores have generated less consistent results. There is an emerging view from research in New Zealand that divaricate branching evolved to deter giant browsing birds (moa). However, alternative explanations, including physiological and climate-based mechanisms, cannot yet be ruled out. Although divaricate branching is widespread in the Malagasy flora, it does not appear to be widespread on other toothless herbivore islands, such as Hawai'i. Until an overarching test for an increase in the incidence of divaricate branching on toothless browsers islands is conducted, we will not know whether it is a component of the island syndrome in plants.

Spinescence is often argued to be absent from isolated islands, including archipelagos that once housed toothless browsers. Upon closer inspection, spinescence does occur in many island floras. However, it is often deployed differently on toothless browser islands than on continents, often along the margins of leaves and other edible parts, which may have made them more difficult for toothless browsers to swallow whole.

Leaf heteroblasty is widespread on several toothless browser islands. Large numbers of plant species in New Caledonia, New Zealand, and Mauritius undergo metamorphic changes in leaf morphology during ontogeny. In many cases, these changes are consistent with how leaf size, toughness, colour, and spinescence might deter toothless herbivores foraging for individual leaves using a plucking motion. Many of these attributes are lost on smaller islands surrounding New Zealand, which is consistent with the hypothesis that they have evolved to deter avian herbivory. Nevertheless, the precise adaptive significance of attributes associated with leaf hetero-blasty have yet to be established definitively, so whether convergence in plant defensive adaptations against toothless herbivores is a component of an island syndrome in plants awaits further study.

Differences in Dispersal

Darwin's loss of dispersibility hypothesis is one of the oldest and perhaps best-known predictions in island evolution. It postulates that island plant populations should evolve reduced dispersibility to avoid 'sea-swept' costs of dispersal. As logical as it may seem, the evidence for the loss of dispersibility isn't convincing. When it is observed, the loss of dispersal potential is typically associated with increased seed size, rather than the loss of functionality of dispersal aides, as implied by Darwin's shipwrecked mariner analogy. This casts serious doubt over whether the loss of dispersal potential is a component of an island syndrome in plants. Instead, seed gigantism appears to be a repeated pattern in evolution on islands, and the loss of dispersal potential, when it occurs, may often arise as a passive by-product of selection for increased seed size. Future tests of the size constraints hypothesis are needed to test whether this is indeed the case.

Strangely, previous research has focused exclusively on just one of the two hypotheses outlined by Darwin (1859) in his shipwrecked mariner analogy. Approximately a third of previous work on the dispersal potential of island populations has uncovered evidence for greater dispersal potential on islands. Could this be a component of an island syndrome in plants?

Reproductive Biology

Extensive analyses of self-compatibility across the globe in several specious plant groups suggest that self-compatibility is over-represented on islands relative to continents. Baker's law therefore appears to be a component of an island syndrome in plants. On the other hand, global analyses of the distribution of dioecy is inconsistent with the hypothesis that dioecy is more prevalent on isolated islands. Instead, dioecy increases with precipitation, suggesting that the perceived relationship between dioecy and insularity is a spurious by-product of its association with precipitation. Dioecy is known to have evolved repeatedly within many archipelagos, which is consistent with Baker's more complete view of how plant reproductive strategies are shaped by insularity. However, our understanding of *in situ* evolution of dioecy on islands is incomplete. Does immigrant selection initially favour self-compatible colonists, while subsequent natural selection favours outcrossing mechanisms (e.g., dioecy, heterostyly) to avoid inbreeding depression? Answering this question poses a serious challenge for future research.

Although more poorly understood, wind pollination (anemophily) also appears to be over-represented in insular environments. Several factors may be responsible for insular anemophily. It could be selected for by pollinator limitation, given the depauperate nature of island pollinator communities. Alternatively, it could be a more efficient mode of pollination given that islands tend to be windier places than continents. Evidence to date also indicates that flowers differ regularly in both size and shape on islands, with island flowers often being smaller and more generalised than their continental counterparts.

Size Changes

Nearly all available data indicate that island plants produce bigger seeds and leaves relative to their mainland relatives. Seeds in particular provide a remarkable example of convergent evolution in insular gigantism. Similar patterns were observed in leaf size, suggesting that leaf gigantism is also a component of an island syndrome in plants. Nevertheless, next to nothing is known about why evolution might favour leaf and seed gigantism on islands.

Unlike seed and leaf sizes, previous research on the stature of island plants has failed to uncover evidence for convergent size changes. Instead, evidence for both dwarfism and gigantism has been observed. Consistent with the island rule, several new analyses have linked changes in stature to stature itself. Small-statured plants on islands surrounding New Zealand tended to increase in size, while large-statured plants declined. However, with only a single test from one of the world's many archipelagos to go by, it can hardly be considered a repeated pattern in island evolution at present.

Insular woodiness has intrigued island biologists for decades. However, several issues need to be overcome before a firm conclusion can be made over whether it is a repeated pattern in island evolution. First, comparisons of the frequency of insular woodiness relative to secondary woodiness on continents is needed to establish whether the weeds-to-trees pathway occurs more frequently on islands. Second, many of the lineages chosen for previous study of insular woodiness were selected because it was suspected that woodiness had evolved *in situ*, which might inflate the evidence for insular woodiness. Lastly, although dwarfism is known to occur in many plant lineages, few studies have explored whether this results in the loss of secondary woodiness on islands. Does it ever occur?

Loss of Fire-Adapted Traits

Given their maritime climates and non-contiguous geography, fire tends to be less important ecologically on islands than it is on continents. Although previous work is limited, much of it illustrates the evolutionary loss of fire-adapted traits, including the loss of lignotubers, reductions in bark thickness, and the loss of serotinous fruits or cones. The loss of fire-adapted traits is a previously unappreciated trend in island evolution, which highlights the possibility that other trends have gone unnoticed and await discovery.

AN ISLAND SYNDROME IN PLANTS

To do science is to search for repeated patterns

Robert MacArthur (1972)

What can we conclude from our search for an island syndrome in plants? Four trends seem to pass the test: defence displacement, Baker's law, seed gigantism, and leaf gigantism. Ten other life history attributes of island plants appear to have evolved repeatedly on islands, but not enough is yet known to confirm their status as components of an island syndrome. Conversely, two attributes that have traditionally been considered key components of an island syndrome, the loss of dispersibility and insular dioecy, do not appear to have evolved repeatedly on islands.

The search for an island syndrome in plants need not be restricted to the 16 life history attributes listed in Table 7.1. Many other attributes could evolve repeatedly on islands and are well worth future study. Hybridisation, enhanced population variation, changes in niche breadth, and the evolution of mutualistic 'super-generalists' were not covered here, but could very well be repeated patterns in the evolution of island plants (see Grant 1998; Whittaker & Fernández-Palacios 2007). Species interactions on islands, both between plants and animals (e.g., Burns 2013b, Schleuning et al. 2014; Biddick & Burns 2018), and between different types of plants (Burns 2007; Burns & Zotz 2010), are typically viewed as 'networks' (Bascompte & Jordano 2013). Island networks have been the subject of intense research effort, and, although not a life history attribute, plant interaction networks may repeatedly evolve similar topologies on islands (e.g., Traveset et al. 2016b).

The lack of empirical support for the loss of dispersibility and insular dioecy has important implications for the future of the discipline philosophically. They illustrate that putative patterns in island evolution can be falsified, and, if they can be falsified, the discipline is not a scientific dead-end of 'just so' natural history stories. Moving into the future, insightful observations of natural historians like Sherwin Carlquist are best viewed as hypotheses, subject to repeated empirical testing. Should a hypothesised trend not hold up to quantitative testing, it should be abandoned in favour of new, amended predictions of how insular selection pressures favour the evolution of island plants.

Table 7.1 *A summary of the evidence for an island syndrome in plants. Two life history attributes that have traditionally been considered to be repeated patterns in island evolution, but now do not appear to be a component of an island syndrome in plants, are shown on the left ('doubtful components'). Ten possible components are listed in the middle ('possible components'). Four apparently repeated patterns in island plants are shown on the right ('probable components')*

Doubtful components:	Possible components	Probable components
1. Loss of dispersibility	1. Leaf heteroblasty and spinescence	1. Defence displacement
2. Insular dieocy	2. Loss of protection mutualisms	2. Baker's law
	3. Divaricate branching	3. Seed gigantism
	4. Seed-size constraints on dispersal	4. Leaf gigantism
	5. Island rule in plant stature	
	6. Insular woodiness	
	7. Reduced incidence of heterostyly	
	8. Anemophily	
	9. Generalised floral morphology	
	10. Loss of fire-adapted traits	

The search for an island syndrome in plants can be a rigorous scientific discipline with much to teach the world of science. Identifying general trends in the evolution of plant traits on islands not only has the potential to generate valuable new insight into plant form and function, but many of the life history trends covered in the proceeding chapters also appear to have parallels in the animal kingdom. This hints at the possibility of a unified search for joint island syndrome in both plants and animals (see Whittaker et al. 2017), as insular selection pressures may well shape all life on earth in similar ways.

References

Ackerly, D D, & Donoghue, M J. (1998). Leaf size, sapling allometry, and Corner's rules: phylogeny and correlated evolution in maples (*Acer*). *American Naturalist*, **152**, 767–791.

Adsersen, A, & Adsersen, H. (1993). Cyanogenic plants in the Galápagos Islands: ecological and evolutionary aspects. *Oikos*, **67**, 511–520.

Adsersen, H. (1995). Research on islands: classic, recent, and prospective approaches. In *Islands* (pp. 7–21). Berlin, Heidelberg: Springer.

Anderson, G J, & Bernardello, G. (2018). Reproductive biology. In: *Plants of Oceanic Islands: Evolution, Biogeography, and Conservation of the Flora of the Juan Fernández (Robinson Crusoe Islands)*. Eds: Stuessy, T F, Crawford, D J, Lopez-Sepulveda, P, Baeza, C M, & Ruiz, E A. Cambridge: Cambridge University Press.

Anderson, G J, Bernardello, G, Stuessy, T F, & Crawford, D J. (2001). Breeding system and pollination of selected plants endemic to Juan Fernández Islands. *American Journal of Botany*, **88**, 220–233.

Agenbroad, L D. (2010). Mammuthus exilis from the California Channel Islands: Height, mass and geologic age. *Proceedings of the 7th California Islands Symposium*, **173**, 536.

Agrawal, A A, & Weber, M G. (2015). On the study of plant defence and herbivory using comparative approaches: how important are secondary plant compounds. *Ecology Letters*, **18**, 985–991.

Allan, H H. (1961). *Flora of New Zealand*. Volume I. Wellington: Government Printer.

Allan, H H. (1982) *Flora of New Zealand*. Volume I. Wellington: Government Printer.

Archer, C L, & Jacobson, M Z. (2005). Evaluation of global wind power. *Journal of Geophysical Research: Atmospheres*, **110**, D12110.

Archibald, S, Lehmann, C E, Belcher, C M, et al. (2018). Biological and geophysical feedbacks with fire in the Earth system. *Environmental Research Letters*, **13**, 033003.

Argue, D, Donlon, D, Groves, C, & Wright, R. (2006). Homo floresiensis: microcephalic, pygmoid, Australopithecus, or Homo? *Journal of Human Evolution*, **51**, 360–374.

Argue, D, Groves, C P, Lee, M S, & Jungers, W L. (2017). The affinities of Homo floresiensis based on phylogenetic analyses of cranial, dental, and postcranial characters. *Journal of Human Evolution*, **107**, 107–133.

Armbruster, W S, Pélabon, C, Bolstad, G H, & Hansen, T F. (2014). Integrated phenotypes: understanding trait covariation in plants and animals. *Philosophical Transactions of the Royal Society of London, B*, **369**, 20130245.

Atkinson, I A, & Greenwood, R M. (1989). Relationships between moas and plants. *New Zealand Journal of Ecology*, **12**, 67–96.

Attard, M R, Wilson, L A, Worthy, T H, et al. (2016). Moa diet fits the bill: virtual reconstruction incorporating mummified remains and prediction of biomechanical performance in avian giants. *Proceedings of the Royal Society of London, B*, **283**, 2015–2043.

Austin, J J, Arnold, E N, & Bour, R. (2003). Was there a second adaptive radiation of giant tortoises in the Indian Ocean? Using mitochondrial DNA to investigate speciation and biogeography of *Aldabrachelys* (Reptilia, Testudinidae). *Molecular Ecology*, **12**, 1415–1424.

Azzani, L. (2015) The origin and function of odours in island birds. Doctoral thesis, University of Canterbury, New Zealand.

Baker, H G. (1955). Self-compatibility and establishment after 'long-distance' dispersal. *Evolution*, **9**, 347–349.

Baker, H G. (1967). Support for Baker's law – as a rule. *Evolution*, **21**, 853–856.

Baker, H G, & Cox, P A. (1984). Further thoughts on dioecism and islands. *Annals of the Missouri Botanical Garden*, **71**, 244–253.

Baldwin, B G. (2007). Adaptive radiation of shrubby tarweeds (Deinandra) in the California Islands parallels diversification of the Hawaiian silversword alliance (Compositae–Madiinae). *American Journal of Botany*, **94**, 237–248.

Baldwin, B G, & Sanderson, M J. (1998). Age and rate of diversification of the Hawaiian silversword alliance (Compositae). *Proceedings of the National Academy of Sciences*, **95**, 9402–9406.

Baldwin, B G, & Wagner, W L. (2010). Hawaiian angiosperm radiations of North American origin. *Annals of Botany*, **105**, 849–879.

Balance, A. (2001). Takahe: The bird that twice came back from the grave. In: *The Takahe: Fifty Years of Conservation Management and Research*. Eds: Lee, W G & Jamieson, I G. Dunedin: University of Otago Press.

Barber, J C, Francisco-Ortega, J, Santos-Guerra, A, Turner, K G, & Jansen, R K. (2002). Origin of Macaronesian *Sideritis* L. (Lamioideae: Lamiaceae) inferred from nuclear and chloroplast sequence datasets. *Molecular Phylogenetics and Evolution*, **23**, 293–306.

Barlow, B A, & Wiens, D. (1977). Host-parasite resemblance in Australian mistletoes: the case for cryptic mimicry. *Evolution*, **31**, 69–84.

Barton, K E. (2014). Prickles, latex, and tolerance in the endemic Hawaiian prickly poppy (Argemone glauca): variation between populations, across ontogeny, and in response to abiotic factors. *Oecologia*, **174**, 1273–1281.

Barton, K E. (2016). Tougher and thornier: general patterns in the induction of physical defence traits. *Functional Ecology*, **30**, 181–187.

Barrett, S C H. (1996). The reproductive biology and genetics of island plants. *Philosophical Transactions of the Royal Society of London, Series B*, **351**, 725–733.

Barrett, S C H, & Shore, J S. (2008). New insights on heterostyly: comparative biology, ecology and genetics. In: *Self-Incompatibility in Flowering Plants* (pp. 3–32). Berlin Heidelberg: Springer.

Barrett, S C, & Hough, J. (2012). Sexual dimorphism in flowering plants. *Journal of Experimental Botany*, **64**, 67–82.

Bascompte, J, & Jordano, P. (2013). *Mutualistic Networks*. Volume 70. Princeton, NJ: Princeton University Press.

Battersby, P F, Wilmshurst, J M, Curran, T J, & Perry, G L. (2017a). Does heating stimulate germination in *Leptospermum scoparium* (mānuka; Myrtaceae)? *New Zealand Journal of Botany*, **55**, 452–465.

Battersby, P F, Wilmshurst, J M, Curran, T J, McGlone, M S, & Perry, G L. (2017b). Exploring fire adaptation in a land with little fire: serotiny in *Leptospermum scoparium* (Myrtaceae). *Journal of Biogeography*, **44**, 1306–1318.

Bawa, K S. (1982). Outcrossing and the incidence of dioecism in island floras. *American Naturalist*, **119**, 866–871.

Beattie, A J. (1985). *The Evolutionary Ecology of Ant-Plant Mutualisms*. Cambridge: Cambridge University Press.

Beever, R E. (1986). Large leaves plants of the northern offshore islands, New Zealand. *Offshore Islands of Northern New Zealand, New Zealand Department of Lands and Survey Information Series*, **16**, 51–61.

Benton, M J, Csiki, Z, Grigorescu, D, et al. (2010). Dinosaurs and the island rule: The dwarfed dinosaurs from Haţeg Island. *Palaeogeography, Palaeoclimatology, Palaeoecology*, **293**, 438–454.

Bergeron, Y, & Brisson, J. (1990). Fire regime in red pine stands at the northern limit of the species' range. *Ecology*, **71**, 1352–1364.

Biddick, M, & Burns, K C. (2018). Phenotypic trait matching predicts the topology of an insular plant–bird pollination network. *Integrative Zoology*, **13**, 339–347.

Biddick, M, & Burns, K C. (2019). Independent evolution of allometric traits: A test of the allometric constraint hypothesis in island vines. *Biological Journal of the Linnean Society*, **126**: 203–211.

Biddick, M, Hutton, I, & Burns, K C. (2018). An alternative water transport system in land plants. *Proceedings of the Royal Society B*, **285**, 20180995.

Blackmore, S, Chin, S C, Chong Seng, L, et al. (2012). Observations on the morphology, pollination and cultivation of coco de mer (*Lodoicea maldivica* (JF Gmel.) Pers., Palmae). *Journal of Botany*, **2012**, article ID 687832.

Blick, R A J, Burns, K C, & Moles, A T. (2012). Predicting network topology of mistletoe–host interactions: do mistletoes really mimic their hosts? *Oikos*, **121**, 761–771.

Blumstein, D T. (2002). Moving to suburbia: ontogenetic and evolutionary consequences of life on predator-free islands. *Journal of Biogeography*, **29**, 685–692.

Böhle, U R, Hilger, H H, & Martin, W F. (1996). Island colonization and evolution of the insular woody habit in Echium L. (Boraginaceae). *Proceedings of the National Academy of Sciences*, **93**, 11740–11745.

Bond, W J, & Midgley, J J. (2001). Ecology of sprouting in woody plants: the persistence niche. *Trends in Ecology & Evolution*, **16**, 45–51.

Bond, W J, Lee, W G, & Craine, J M. (2004a). Plant structural defences against browsing birds: a legacy of New Zealand's extinct moas. *Oikos*, **104**, 500–508.

Bond, W J, Dickinson, K J, & Mark, A F. (2004b). What limits the spread of fire-dependent vegetation? Evidence from geographic variation of serotiny in a New Zealand shrub. *Global Ecology & Biogeography*, **13**, 115–127.

Bond, W J, & Silander, J A. (2007). Springs and wire plants: anachronistic defences against Madagascar's extinct elephant birds. *Proceedings of the Royal Society of London, B*, **274**, 1985–1992.

Bonte, D, Van Dyck, H, Bullock, J M, et al. (2012). Costs of dispersal. *Biological Reviews*, **87**, 290–312.

Bowen, L, & Van Vuren, D. (1997). Insular endemic plants lack defenses against herbivores. *Conservation Biology*, **11**, 1249–1254.

Bowmaker, J K. (1998). Evolution of colour vision in vertebrates. *Eye*, **12**, 541–547.

Bowman, D M, Balch, J K, Artaxo, P, et al. (2009). Fire in the Earth system. *Science*, **324**, 481–484.

Bradshaw, S D, Dixon, K W, Hopper, S D, Lambers, H, & Turner, S R. (2011). Little evidence for fire-adapted plant traits in Mediterranean climate regions. *Trends in Plant Science*, **16**, 69–76.

Bradstock, R A, Williams, R J, & Gill, A M. (Eds.). (2012). *Flammable Australia: Fire Regimes, Biodiversity and Ecosystems in a Changing World*. Clayton: CSIRO Publishing.

Bramow, C, Hartvig, I, Larsen, S B, & Philipp, M. (2013). How a heterostylous plant species responds to life on remote islands: a comparative study of the morphology and reproductive biology of *Waltheria ovata* on the coasts of Ecuador and the Galápagos Islands. *Evolutionary Ecology*, **27**, 83–100.

Briand, C H, Schwilk, D W, Gauthier, S, & Bergeron, Y. (2015). Does fire regime influence life history traits of jack pine in the southern boreal forest of Québec, Canada? *Plant Ecology*, **216**, 157–164.

Brokaw, N V. (1985). Gap-phase regeneration in a tropical forest. *Ecology*, **66**, 682–687.

Bromham, L, & Cardillo, M. (2007). Primates follow the 'island rule': implications for interpreting Homo floresiensis. *Biology letters*, **3**, 398–400.

Brown, P, Sutikna, T, Morwood, M J, Soejono, R P, Saptomo, E W, & Due, R A. (2004). A new small-bodied hominin from the Late Pleistocene of Flores, Indonesia. *Nature*, **431**, 1055.

Brown, V K, & Lawton, J H. (1991). Herbivory and the evolution of leaf size and shape. *Philosophical Transactions of the Royal Society of London, B*, **333**, 265–272.

Brumm, A, Van den Bergh, G D, Storey, M, et al. (2016). Age and context of the oldest known hominin fossils from Flores. *Nature*, **534**, 249.

Bryant, J P, Tahvanainen, J, Sulkinoja, M, Julkunen-Tiitto, R, Reichardt, P, & Green, T. (1989). Biogeographic evidence for the evolution of chemical defense by boreal birch and willow against mammalian browsing. *American Naturalist*, **134**, 20–34.

Burns, K C. (2002). Seed dispersal facilitation and geographic consistency in bird–fruit abundance patterns. *Global Ecology & Biogeography*, **11**, 253–259.

Burns, K C. (2004). Scale and macroecological patterns in seed dispersal mutualisms. *Global Ecology & Biogeography*, **13**, 289–293.

Burns, K C. (2007). Network properties of an epiphyte metacommunity. *Journal of Ecology*, **95**, 1142–1151.

Burns, K C. (2008). When is it coevolution? A reply to Morgan-Richards et al. *New Zealand Journal of Ecology*, **32**, 113.

Burns, K C. (2013a). Are there general patterns in plant defence against megaherbivores? *Biological Journal of the Linnean Society*, **111**, 38–48.

Burns, K C. (2013b). What causes size coupling in fruit–frugivore interaction webs? *Ecology*, **94**, 295–300.

Burns, K C. (2016a). Spinescence in the New Zealand flora: parallels with Australia. *New Zealand Journal of Botany*, **54**, 273–289.

Burns, K C. (2016b). Size changes in island plants: independent trait evolution in *Alyxia ruscifolia* (Apocynaceae) on Lord Howe Island. *Biological Journal of the Linnean Society*, **119**, 847–855.

Burns, K C. (2018). Time to abandon the loss of dispersal ability hypothesis in island plants: A comment on García-Verdugo, Mairal, Monroy, Sajeva and Caujapé-Castells (2017). *Journal of Biogeography*, **45**, 1219–1222.

Burns, K C, & Dawson, J W. (2006). A morphological comparison of leaf hetero-blasty between New Caledonia and New Zealand. *New Zealand Journal of Botany*, **44**, 387–396.

Burns, K C, & Dawson, J W. (2009). Heteroblasty on Chatham Island: a comparison with New Zealand and New Caledonia. *New Zealand Journal of Ecology*, **33**, 156–163.

Burns, K C, & Zotz, G. (2010). A hierarchical framework for investigating epiphyte assemblages: networks, meta-communities, and scale. *Ecology*, **91**, 377–385.

Burns, K C, Herold, N, & Wallace, B. (2012). Evolutionary size changes in plants of the south-west Pacific. *Global Ecology & Biogeography*, **21**, 819–828.

Caccone, A, Amato, G, Gratry, O C, Behler, J, & Powell, J R. (1999). A molecular phylogeny of four endangered Madagascar tortoises based on mtDNA sequences. *Molecular Phylogenetics and Evolution*, **12**, 1–9.

Caccone, A, Gentile, G, Gibbs, J P, et al. (2002). Phylogeography and history of giant Galápagos tortoises. *Evolution*, **56**, 2052–2066.

Carlquist, S. (1966a). The biota of long-distance dispersal. II. Loss of dispersibility in Pacific Compositae. *Evolution*, **20**, 30–48.

Carlquist, S. (1966b) The biota of long-distance dispersal. I. Principles of dispersal and evolution. *Quarterly Reviews in Biology*, **41**, 247–270.

Carlquist, S. (1966c). The biota of long-distance dispersal. III. Loss of dispersibility in the Hawaiian flora. *Brittonia*, **18**, 310–335.

Carlquist, S. (1974). *Island Biology*. New York, London: Columbia University Press.

Carlquist, S. (1976). Alexgeorgea, a bizarre new genus of Restionaceae from Western Australia. *Australian Journal of Botany*, **24**, 281–295.

Carlquist, S. (1980). *Hawaii: A Natural History*. Garden City, NY: Natural History Press.

Carlquist, S. (2001). Wood anatomy of the endemic woody Asteraceae of St Helena I: phyletic and ecological aspects. *Botanical Journal of the Linnean Society*, **137**, 197–210.

Carlquist, S. (2009). Darwin on island plants. *Botanical Journal of the Linnean Society*, **161**, 20–25.

Carroll, M C, Laughrin, L L, & Bromfield, A C. (1993). Fire on the California Islands: Does it play a role in chaparral and closed cone pine forest habitats? In: *Recent Advances and Research on the California Islands* (pp. 73–88). Santa Barbara, CA: Santa Barbara Museum of Natural History.

Carvajal-Endara, S, Hendry, A P, Emery, N C, & Davies, T J. (2017). Habitat filtering not dispersal limitation shapes oceanic island floras: species assembly of the Galápagos archipelago. *Ecology Letters*, **20**, 495–504.

Cash, V W, & Fulbright, T E. (2005). Nutrient enrichment, tannins, and thorns: effects on browsing of shrub seedlings. *Journal of Wildlife Management*, **69**, 782–793.

Causley, C L, Fowler, W M, Lamont, B B, & He, T. (2016). Fitness benefits of serotiny in fire-and drought-prone environments. *Plant Ecology*, **217**, 773–779.

Charles-Dominique, T, Barczi, J F, Le Roux, E, & Chamaillé-Jammes, S. (2017). The architectural design of trees protects them against large herbivores. *Functional Ecology*, **31**, 1710–1717.

Charnov, E L. (1979). Simultaneous hermaphroditism and sexual selection. *Proceedings of the National Academy of Sciences*, **76**, 2480–2484.

Chase, J M, & Leibold, M A. (2003). *Ecological Niches: Linking Classical and Contemporary Approaches*. Chicago: University of Chicago Press.

Cheke, A S, & Hume, J. (2008). *Lost Land of the Dodo*. London: T & A Poyser.

Cheptou, P O, Carrue, O, Rouifed, S, & Cantarel, A. (2008). Rapid evolution of seed dispersal in an urban environment in the weed *Crepis sancta*. *Proceedings of the National Academy of Sciences*, **105**, 3796–3799.

Chiari, Y, Meijden, A, Caccone, A, Claude, J, & Gilles, B. (2017). Self-righting potential and the evolution of shell shape in Galápagos tortoises. *Scientific Reports*, **7**, 15828.

Christian, R, Kelly, D, & Turnbull, M H. (2006). The architecture of New Zealand's divaricate shrubs in relation to light adaptation. *New Zealand Journal of Botany*, **44**, 171–186.

Chuvieco, E, Giglio, L, & Justice, C. (2008). Global characterization of fire activity: toward defining fire regimes from Earth observation data. *Global Change Biology*, **14**, 1488–1502.

Clarke, P J, Lawes, M J, Murphy, B P, et al. (2015). A synthesis of postfire recovery traits of woody plants in Australian ecosystems. *Science of the Total Environment*, **534**, 31–42.

Climent, J, Tapias, R, Pardos, J A, & Gil, L. (2004). Fire adaptations in the Canary Islands pine (*Pinus canariensis*). *Plant Ecology*, **171**, 185–196.

Cody, M L. (1984). Branching patterns in columnar cacti. In: *Being Alive on Land* (pp. 201–236). Eds: Margaris, N, Arianoutsou-Farragitaki, M, & Oechel, W. Dordrecht: Springer.

Cody, M L. (2002). Plants. In: *A New Island Biogeography of the Sea of Cortez* (pp. 63–111). Eds: Case, T J, Cody, M L, & Ezcurra, E. Oxford: Oxford University Press.

Cody, M L. (2006). *Plants on Islands: Diversity and Dynamics on a Continental Archipelago*. Berkeley: University of California Press.

Cody, M L, & McC Overton, J. (1996). Short-term evolution of reduced dispersal in island plant populations. *Journal of Ecology*, **84**, 53–61.

Colyvan, M, & Ginzburg, L R. (2003). Laws of nature and laws of ecology. *Oikos*, **101**, 649–653.

Connell, J H. (1971). On the role of natural enemies in preventing competitive exclusion in some marine animals and in rain forest trees. In: *Dynamics of Populations* (pp. 298–312). Eds: Den Boer, P J & Gradwell, G. Oosterbeek: PUDOC.

Coomes, D A, & Grubb, P J. (2003). Colonization, tolerance, competition and seed-size variation within functional groups. *Trends in Ecology & Evolution*, **18**, 283–291.

Cornelissen, J H C, Lavorel, S, Garnier, E, et al. (2003). A handbook of protocols for standardised and easy measurement of plant functional traits worldwide. *Australian Journal of Botany*, **51**, 335–380.

Corner, E J H. (1949). The durian theory or the origin of the modern tree. *Annals of Botany*, **13**, 367–414.

Cox, B T M, & Burns, K C. (2017). Convergent evolution of gigantism in the flora of an isolated archipelago. *Evolutionary Ecology*, **31**, 741–752.

Coverdale, T C, Goheen, J R, Palmer, T M, & Pringle, R M. (2018). Good neighbors make good defenses: associational refuges reduce defense investment in African savanna plants. *Ecology*, **99**, 1724–1736.

Crawford, D J, Anderson, G J, & Bernardello, G. (2011). The reproductive biology of island plants. In: *The Biology of Island Floras*. Bramwell, D & Caujape-Castells, J. Cambridge: Cambridge University Press.

Cronk, Q C B. (1992). Relict floras of Atlantic islands: patterns assessed. *Biological Journal of the Linnean Society*, **46**, 91–103.

Crosti, R, Ladd, P G, Dixon, K W, & Piotto, B. (2006). Post-fire germination: The effect of smoke on seeds of selected species from the central Mediterranean basin. *Forest Ecology and Management*, **221**, 306–312.

Crowley, B E, & Godfrey, L R. (2013). Why all those spines? Anachronistic defences in the Didiereoideae against now extinct lemurs. *South African Journal of Science*, **109**, 1–7.

Culek, M. (2013). Geological and morphological evolution of the Socotra Archipelago (Yemen) from the biogeographical view. *Journal of Landscape Ecology*, **6**, 84–108

Darlington, P J. (1957). *Zoogeography: The Geographical Distribution of Animals*. New York: Wiley.

Darwin, C R. (1859). *On the Origin of Species by Means of Natural Selection, or the Preservation of Favoured Races in the Struggle for Life*. 1st edition. London: John Murray.

Danell, K, & Bergström, R. (2002). Mammalian herbivory in terrestrial environments. In: *Plant-Animal Interacts, an Evolutionary Approach* (pp. 107–131). Eds: Herrera, C M & Pellmyr, O. Oxford: Blackwell Publishing.

Dawson, E Y. (1966). Cacti on the Galapagos Islands, with special reference to their relations with tortoises. In: *The Galapagos* (pp. 209–214). Ed: Bowman, R I. Berkeley: University of California Press.

Dawson, J. (1988). *Forest Vines to Snow Tussocks: The Story of New Zealand Plants*. Wellington: Victoria University Press.

Day, J S. (1998). Light conditions and the evolution of heteroblasty (and the divaricate form) in New Zealand. *New Zealand Journal of Ecology*, **22**, 43–54.

Day, J S, & Gould, K S. (1997). Vegetative architecture of *Elaeocarpus hookerianus*. Periodic growth patterns in divaricating juveniles. *Annals of Botany*, **79**, 607–616.

Day, J S, Gould, K S, & Jameson, P E. (1997). Vegetative architecture of *Elaeocarpus hookerianus*. Transition from juvenile to adult. *Annals of Botany*, **79**, 617–624.

Dell'Aglio, D D, Losada, M E, & Jiggins, C D. (2016). Butterfly learning and the diversification of plant leaf shape. *Frontiers in Ecology and Evolution*, **4**, 81.

de Waal, C, Rodger, J G, Anderson, B, & Ellis, A G. (2014). Selfing ability and dispersal are positively related, but not affected by range position: a multi-species study on southern African Asteraceae. *Journal of Evolutionary Biology*, **27**, 950–959.

Díaz Vélez, M C, Ferreras, A E, Silva, W R, & Galetto, L. (2017). Does avian gut passage favour seed germination of woody species of the Chaco Serrano Woodland in Argentina? *Botany*, **95**, 493–501.

Dulin, M W, & Kirchoff, B K. (2010). Paedomorphosis, secondary woodiness, and insular woodiness in plants. *Botanical Review*, **76**, 405–490.

Duncan, R P, Boyer, A G, & Blackburn, T M. (2013). Magnitude and variation of prehistoric bird extinctions in the Pacific. *Proceedings of the National Academy of Sciences*, **110**, 6436–6441.

Duthie, C, Gibbs, G, & Burns, K C. (2006). Seed dispersal by weta. *Science*, **311**, 1575–1575.

Edwards, P J, Kollmann, J, & Fleischmann, K. (2002). Life history evolution in *Lodoicea maldivica* (Arecaceae). *Nordic Journal of Botany*, **22**, 227–238.

Edwards, P J, Fleischer-Dogley, F, & Kaiser-Bunbury, C N. (2015). The nutrient economy of *Lodoicea maldivica*, a monodominant palm producing the world's largest seed. *New Phytologist*, **206**, 990–999.

Edwards, R D, Craven, L A, Crisp, M D, & Cook, L G. (2010). Melaleuca revisited: cpDNA and morphological data confirm that *Melaleuca* L. (Myrtaceae) is not monophyletic. *Taxon*, **59**, 744–754.

Economo, E P, Sarnat, E M, Janda, M, et al. (2015). Breaking out of biogeographical modules: range expansion and taxon cycles in the hyperdiverse ant genus *Pheidole*. *Journal of Biogeography*, **42**, 2289–2301.

Eskildsen, L I, Olesen, J M, & Jones, C G. (2004). Feeding response of the Aldabra giant tortoise (*Geochelone gigantea*) to island plants showing heterophylly. *Journal of Biogeography*, **31**, 1785–1790.

Evans, A R, Daly, E S, Catlett, K K, et al. (2016). A simple rule governs the evolution and development of hominin tooth size. *Nature*, **530**, 477.

Fadzly, N, & Burns, K C. (2010). Hiding from the ghost of herbivory past: Evidence for crypsis in an insular tree species. *International Journal of Plant Sciences*, **171**, 828–833.

Fadzly, N, Jack, C, Schaefer, H M, & Burns, K C. (2009). Ontogenetic colour changes in an insular tree species: signalling to extinct browsing birds? *New Phytologist*, **184**, 495–501.

Falk, D, Hildebolt, C, Smith, K, et al. (2005). The brain of LB1, *Homo floresiensis*. *Science*, **308**, 242–245.

Farmer, E E. (2014). *Leaf Defence*. Oxford: Oxford University Press.

Faurby, S, & Svenning, J C. (2016). Resurrection of the island rule: Human-driven extinctions have obscured a basic evolutionary pattern. *American Naturalist*, **187**, 812–820.

Felger, R S, & Wilder, B T. (2012). *Plant Life of a Desert Archipelago*. Tucson: University of Arizona Press.

Fester, K. (2010). Plant Alkaloids. In *Encyclopedia of Life Sciences*. Chichester: John Wiley and Sons.

Fitzpatrick, J W. (2001). Forward. In *The Takahe: Fifty Years of Conservation Management and Research* (pp. 9–10). Eds: Lee, W G & Jamieson, I G Dunedin: University of Otago Press.

Fleischer-Dogley, F, Kettle, C J, Edwards, P J, Ghazoul, J, Määttänen, K, & Kaiser-Bunbury, C N. (2011). Morphological and genetic differentiation in populations of the dispersal-limited coco de mer (*Lodoicea maldivica*): implications for management and conservation. *Diversity and Distributions*, **17**, 235–243.

Fleming, T H, Maurice, S, Buchmann, S L, & Tuttle, M D. (1994). Reproductive biology and relative male and female fitness in a trioecious cactus, *Pachycereus pringlei* (Cactaceae). *American Journal of Botany*, **81**, 858–867.

Fleming, T H, Maurice, S, & Hamrick, J L. (1998). Geographic variation in the breeding system and the evolutionary stability of trioecy in *Pachycereus pringlei* (Cactaceae). *Evolutionary Ecology*, **12**, 279–289.

Foster, J B. (1964). Evolution of mammals on islands. *Nature*, **202**, 234–235.

Fresnillo, B, & Ehlers, B K. (2008). Variation in dispersability among mainland and island populations of three wind dispersed plant species. *Plant Systematics & Evolution*, **270**, 243–255.

Friedman, J, & Barrett, S C. (2009). Wind of change: New insights on the ecology and evolution of pollination and mating in wind-pollinated plants. *Annals of Botany*, **103**, 1515–1527.

Friedmann, F, & Cadet, T. (1976). Observations sur l'heterophyllie dans les Iles Mascareignes. *Adansonia*, **15**, 423–440.

Fuertes-Aguilar, J, Ray, M F, Francisco-Ortega, J, Santos-Guerra, A, & Jansen, R K. (2002). Molecular evidence from chloroplast and nuclear markers for multiple colonizations of *Lavatera* (Malvaceae) in the Canary Islands. *Systematic Botany*, **27**, 74–83.

Gaffney, E S. (1996). The postcranial morphology of Meiolania platyceps and a review of the Meiolaniidae. *Bulletin of the AMNH*; no. **229**, 1–166.

Gallaher, T, Callmander, M W, Buerki, S, & Keeley, S C. (2015). A long distance dispersal hypothesis for the Pandanaceae and the origins of the *Pandanus tectorius* complex. *Molecular Phylogenetics and Evolution*, **83**, 20–32.

Gamage, H K, & Jesson, L. (2007). Leaf heteroblasty is not an adaptation to shade: seedling anatomical and physiological responses to light. *New Zealand Journal of Ecology*, **31**, 245–254.

Garcia-R, J C, & Trewick, S A. (2015). Dispersal and speciation in purple swamphens (Rallidae: *Porphyrio*). *Auk*, **132**, 140–155.

García-Verdugo, C, Forest, A D, Balaguer, L, Fay, M F, & Vargas, P. (2010a). Parallel evolution of insular *Olea europaea* subspecies based on geographical structuring of plastid DNA variation and phenotypic similarity in leaf traits. *Botanical Journal of the Linnean Society*, **162**, 54–63.

García-Verdugo, C, Méndez, M, Velázquez-Rosas, N, & Balaguer, L. (2010b). Contrasting patterns of morphological and physiological differentiation across insular environments: phenotypic variation and heritability of light-related traits in *Olea europaea*. *Oecologia*, **164**, 647–655.

García-Verdugo, C, Mairal, M, Monroy, P, Sajeva, M, & Caujapé-Castells, J. (2017). The loss of dispersal on islands hypothesis revisited: Implementing phylogeography to investigate evolution of dispersal traits in *Periploca* (Apocynaceae). *Journal of Biogeography*, **44**, 2595–2606.

Garcillán, P P. (2010). Seed release without fire in *Callitropsis guadalupensis*, a serotinous cypress of a Mediterranean-climate oceanic island. *Journal of Arid Environments*, **74**, 512–515.

Gaston, K J, Chown, S L, & Evans, K L. (2008). Ecogeographical rules: Elements of a synthesis. *Journal of Biogeography*, **35**, 483–500.

Gauthier, S, Bergeron, Y, & Simon, J P. (1996). Effects of fire regime on the serotiny level of jack pine. *Journal of Ecology*, **84**, 539–548.

Geritz, S A, van der Meijden, E, & Metz, J A. (1999). Evolutionary dynamics of seed size and seedling competitive ability. *Theoretical Population Biology*, **55**, 324–343.

Ghebreab, W. (1998). Tectonics of the Red Sea region reassessed. *Earth-Science Reviews*, **45**, 1–44.

Ghiselin, M T. (1969). The evolution of hermaphroditism among animals. *The Quarterly Review of Biology*, **44**, 189–208.

Gianoli, E, & Carrasco-Urra, F. (2014). Leaf mimicry in a climbing plant protects against herbivory. *Current Biology*, **24**, 984–987.

Gibbs, G. (2006) *Ghosts of Gandwana*. Nelson: Craig Potton Publishing.

Givnish, T J. (1997). Adaptive radiation and molecular systematics: aims and conceptual issues. In *Molecular evolution and adaptive radiation* (pp. 1–54). Eds: Givnish, T J & Systma, K J. New York, NY, USA: Cambridge University Press.

Givnish, T J. (1998). Adaptive plant evolution on islands: classical patterns, molecular data, new insights. In *Evolution on Islands* (pp. 281–304). Eds: Grant, P R. Oxford: Oxford University Press.

Givnish, T J. (2015). Adaptive radiation versus 'radiation' and 'explosive diversification': why conceptual distinctions are fundamental to understanding evolution. *New Phytologist*, **207**, 297–303.

Givnish, T J, Millam, K C, Theim, T T, Mast, A R, Patterson, T B, Hipp, A L, Henss, J M, Smith, J F, Wood, K R, Sytsma, K J. (2009). Origin, adaptive radiation, and diversification of the Hawaiian lobeliads (Asterales: Campanulaceae). *Proceedings of the Royal Society B: Biological Sciences*, **276**, 407–416.

Givnish, T J, Sytsma, K J, Smith, J F, & Hahn, W J. (1994). Thorn-like prickles and heterophylly in Cyanea: adaptations to extinct avian browsers on Hawaii? *Proceedings of the National Academy of Sciences*, **91**, 2810–2814.

Gollner, A L. (1999). *The Fruit Hunters, a Story of Nature, Adventure, Commerce and Obsession*. New York: Scribner. ISBN 978-0-7432-9694-6.

Golonka, A M, Sakai, A K, & Weller, S G. (2005). Wind pollination, sexual dimorphism, and changes in floral traits of *Schiedea* (Caryophyllaceae). *American Journal of Botany*, **92**, 1492–1502.

Gordon, A D, Nevell, L, & Wood, B. (2008). The *Homo floresiensis* cranium (LB1): size, scaling, and early Homo affinities. *Proceedings of the National Academy of Sciences*, **105**, 4650–4655.

Goodson, B E, Santos-Guerra, A, & Jansen, R K. (2006). Molecular systematics of *Descurainia* (Brassicaceae) in the Canary Islands: Biogeographic and taxonomic implications. *Taxon*, **55**, 671–682.

Goodman, S M, & Jungers, W L. (2014). *Extinct Madagascar: Picturing the Island's Past*. Chicago: University of Chicago Press.

Göldel, B, Araujo, A C, Kissling, W D, & Svenning, J C. (2016). Impacts of large herbivores on spinescence and abundance of palms in the Pantanal, Brazil. *Botanical Journal of the Linnean Society*, **182**, 465–479.

Gómez, J M, & Zamora, R. (2002). Thorns as induced mechanical defense in a long-lived shrub (*Hormathophylla spinosa*, Cruciferae). *Ecology*, **83**, 885–890.

Grant, P R. (1998). Patterns on islands and microevolution. In: *Evolution on Islands* (pp. 1–17). Eds: Grant, P R. Oxford: Oxford University Press.

Green, P S. (1970) *Sophora howinsula*. *Journal of the Arnold Arboretum*. **51**, 204.

Greene, D F, & Johnson, E A. (1993). Seed mass and dispersal capacity in wind-dispersed diaspores. *Oikos*, **67**, 69–74.

Greenwood, R M, & Atkinson, I A E. (1977). Evolution of divaricating plants in New Zealand in relation to moa browsing. *Proceedings of the New Zealand Ecological Society*, **24**, 21–33.

Grossenbacher, D L, Brandvain, Y, Auld, J R, et al. (2017). Self-compatibility is over-represented on islands. *New Phytologist*, **215**, 469–478.

Grubb, P J. (1977). The maintenance of species-richness in plant communities: the importance of the regeneration niche. *Biological Reviews*, **52**, 107–145.

Grubb, P J. (2003). Interpreting some outstanding features of the flora and vegetation of Madagascar. *Perspectives in Plant Ecology, Evolution & Systematics*, **6**, 125–146.

Grueber, C E, & Jamieson, I G. (2011). Low genetic diversity and small population size of Takahe *Porphyrio hochstetteri* on European arrival in New Zealand. *Ibis*, **153**, 384–394.

Guppy, H B (1906) *Observations of a Naturalist in the Pacific, 2, Plant Dispersal*. MacMillan: London.

Gutiérrez-Flores, C, García-De León, F J, & Cota-Sánchez, J H. (2016). Microsatellite genetic diversity and mating systems in the columnar cactus *Pachycereus pringlei* (Cactaceae). *Perspectives in Plant Ecology, Evolution & Systematics*, **22**, 1–10.

Hagelin, J C. (2004). Observations on the olfactory ability of the Kakapo *Strigops habroptilus*, the critically endangered parrot of New Zealand. *Ibis*, **146**, 161–164.

Hamilton-Brown, S, Boon, P I, Raulings, E, Morris, K, & Robinson, R. (2009). Aerial seed storage in *Melaleuca ericifolia* Sm. (Swamp Paperbark): Environmental triggers for seed release. *Hydrobiologia*, **620**, 121–133.

Hanley, M E, Lamont, B B, Fairbanks, M M, & Rafferty, C M. (2007). Plant structural traits and their role in anti-herbivore defence. *Perspectives in Plant Ecology, Evolution & Systematics*, **8**, 157–178.

Hansen, I, Brimer, L, & Mølgaard, P. (2003). Herbivore-deterring secondary compounds in heterophyllous woody species of the Mascarene Islands. *Perspectives in Plant Ecology, Evolution and Systematics*, **6**, 187–203.

Harms, K E, & Dalling, J W. (1997). Damage and herbivory tolerance through resprouting as an advantage of large seed size in tropical trees and lianas. *Journal of Tropical Ecology*, **13**, 617–621.

Harris, W. (2002). Variation of inherent seed capsule splitting in populations of *Leptospermum scoparium* (Myrtaceae) in New Zealand. *New Zealand Journal of Botany*, **40**, 405–417.

Hawkins, S, Worthy, T H, Bedford, S, et al. (2016). Ancient tortoise hunting in the southwest Pacific. *Scientific Reports*, **6**, 38317.

Heenan, P B, Mitchell, A D, De Lange, P J, Keeling, J, & Paterson, A M. (2010). Late-Cenozoic origin and diversification of Chatham Islands endemic plant species revealed by analyses of DNA sequence data. *New Zealand Journal of Botany*, **48**, 83–136.

Herrera, C M. (2002a). Seed dispersal by vertebrates. In: *Plant–Animal Interactions: An Evolutionary Approach* (pp. 185–210). Eds: Herrera, C M & Pellmyr, O. Oxford: Blackwell Science Ltd.

Herrera, C M. (2002b). Correlated evolution of fruit and leaf size in bird-dispersed plants: species-level variance in fruit traits explained a bit further? *Oikos*, **97**, 426–432.

Hesse, E, & Pannell, J R. (2011). Density-dependent pollen limitation and reproductive assurance in a wind-pollinated herb with contrasting sexual systems. *Journal of Ecology*, **99**, 1531–1539.

Hijmans, R J, Cameron, S, Parra, J, Jones, P, Jarvis, A, & Richardson, K. (2008). *WorldClim*. Berkeley: University of California.

Hoan, R P, Ormond, R A, & Barton, K E. (2014). Prickly poppies can get pricklier: ontogenetic patterns in the induction of physical defense traits. *PLoS ONE*, **9**, e96796.

Hochberg, M C. (1980). Factors effecting leaf size of chaparral shrubs on the California Islands. In: *The California Islands* (pp. 189–206). Santa Barbara, CA: Santa Barbara Museum of Natural History.

Hooker, J D. (1853). *Flora Novae-Zelandiae*. Part 1. London: Reeve Brothers.

Howell, C J, Kelly, D, & Turnbull, M H. (2002). Moa ghosts exorcised? New Zealand's divaricate shrubs avoid photoinhibition. *Functional Ecology*, **16**, 232–240.

Huber, M, Triebwasser-Freese, D, Reichelt, M, et al. (2015). Identification, quantification, spatiotemporal distribution and genetic variation of major latex secondary metabolites in the common dandelion (*Taraxacum officinale* agg.). *Phytochemistry*, **115**, 89–98.

Hurr, K A, Lockhart, P J, Heenan, P B, & Penny, D. (1999). Evidence for the recent dispersal of *Sophora* (Leguminosae) around the Southern Oceans: Molecular data. *Journal of Biogeography*, **26**, 565–577.

Hyatt, L A, Rosenberg, M S, Howard, T G, et al. (2003). The distance dependence prediction of the Janzen-Connell hypothesis: a meta-analysis. *Oikos*, **103**, 590–602.

Igic, B, Lande, R, & Kohn, J R. (2008). Loss of self-incompatibility and its evolutionary consequences. *International Journal of Plant Sciences*, **169**, 93–104.

Imbert, E. (1999). The effects of achene dimorphism on the dispersal in time and space in *Crepis sancta* (Asteraceae). *Canadian Journal of Botany*, **77**, 508–513.

Imbert, E, Escarre, J, & Lepart, J. (1996). Achene dimorphism and among-population variation in *Crepis sancta* (Asteraceae). *International Journal of Plant Sciences*, **157**, 309–315.

Inoue, K, & Amano, M. (1986). Evolution of *Campanula punctata* Lam. in the Izu Islands: Changes of pollinators and evolution of breeding systems. *Plant Species Biology*, **1**, 89–97.

Ison, J L, & Wagenius, S. (2014). Both flowering time and distance to conspecific plants affect reproduction in *Echinacea angustifolia*, a common prairie perennial. *Journal of Ecology*, **102**, 920–929.

Itescu, Y, Karraker, N E, Raia, P, Pritchard, P C, & Meiri, S. (2014). Is the island rule general? Turtles disagree. *Global Ecology and Biogeography*, **23**, 689–700.

Jacobs, G H, Deegan, J F, & Neitz, J A Y. (1998). Photopigment basis for dichromatic color vision in cows, goats, and sheep. *Visual Neuroscience*, **15**, 581–584.

Jaffe, A L, Slater, G J, & Alfaro, M E. (2011). The evolution of island gigantism and body size variation in tortoises and turtles. *Biology Letters*, **7**(4):558–561.

James, S. (1984). Lignotubers and burls: their structure, function and ecological significance in Mediterranean ecosystems. *Botanical Review*, **50**, 225–266.

James, H F, & Burney, D A. (1997). The diet and ecology of Hawaii's extinct flightless waterfowl: evidence from coprolites. *Biological Journal of the Linnean Society*, **62**, 279–297.

Jamieson, I G, & Ryan, C J. (2001). Island Takahe: Closure of the debate over the merits of introducing Fiordland Takahe to predator-free islands. In *The Takahe: Fifty Years of Conservation Management and Research* (pp. 96–113). Eds: Lee, W G & Jamieson, I G. Dunedin: University of Otago Press.

Janzen, D H. (1970). Herbivores and the number of tree species in tropical forests. *American Naturalist*, **104**, 501–528.

Janzen, D H. (1973). Dissolution of mutualism between *Cecropia* and its *Azteca* ants. *Biotropica*, **5**, 15–28.

Johnson, D L. (1978). The origin of island mammoths and the quaternary land bridge of the northern Channel Islands, California. *Quaternary Research*, **10**, 204–225.

Johnson, S G, Delph, L F, & Elderkin, C L. (1995). The effect of petal-size manipulation on pollen removal, seed set, and insect-visitor behavior in *Campanula americana*. *Oecologia*, **102**, 174–179.

Jones, C. (2011). Researchers to drill for hobbit history. Nature News, doi:10.1038/news.2011.702

Jønsson, K A, Irestedt, M, Christidis, L, Clegg, S M, Holt, B G, & Fjeldså, J. (2014). Evidence of taxon cycles in an Indo-Pacific passerine bird radiation (Aves: Pachycephala). *Proceedings of the Royal Society of London B: Biological Sciences*, **281**, 20131727.

Kavanagh, P H. (2015). Herbivory and the evolution of divaricate plants: Structural defences lost on an offshore island. *Austral Ecology*, **40**, 206–211.

Kavanagh, P H, & Burns, K C. (2014). The repeated evolution of large seeds on islands. *Proceedings of the Royal Society of London, B*, **281**, 20140675.

Kavanagh, P H, & Burns, K C. (2015). Sexual size dimorphism in island plants: The niche variation hypothesis and insular size changes. *Oikos*, **124**, 717–723.

Kavanagh, P H, Shaw, R C, & Burns, K C. (2016). Potential aposematism in an insular tree species: are signals dishonest early in ontogeny? *Biological Journal of the Linnean Society*, **118**, 951–958.

Keeley, J E. (1987). Role of fire in seed germination of woody taxa in California chaparral. *Ecology*, **68**, 434–443.

Keeley, J E, Pausas, J G, Rundel, P W, Bond, W J, & Bradstock, R A. (2011). Fire as an evolutionary pressure shaping plant traits. *Trends in Plant Science*, **16**, 406–411.

Keeler, K H. (1985). Extrafloral nectaries on plants in communities without ants: Hawaii. *Oikos*, **44**, 407–414.

Kellner, A, Ritz, C M, Schlittenhardt, P, & Hellwig, F H. (2011). Genetic differentiation in the genus *Lithops* L. (Ruschioideae, Aizoaceae) reveals a high level of convergent evolution and reflects geographic distribution. *Plant Biology*, **13**, 368–380.

Kirchman, J J. (2012). Speciation of flightless rails on islands: A DNA-based phylogeny of the typical rails of the Pacific. *Auk*, **129**, 56–69.

Krawchuk, M A, Moritz, M A, Parisien, M A, Van Dorn, J, & Hayhoe, K. (2009). Global pyrogeography: The current and future distribution of wildfire. *PLoS ONE*, **4**, e5102.

Kreiner, J. (2017). Digest: The key to pollen–stigma dimorphisms – dissecting the functional significance of the heterostylous syndrome. *Evolution*, **71**, 187–188.

Krokene, P, Nagy, N E, & Krekling, T. (2010). Traumatic resin duct and polyphenolic parenchyma cells in conifers. In: *Induced Plant Resistance to Herbivory.* Eds: Schaller, A. Berlin: Springer.

Knox, K J E, & Clarke, P J. (2012). Fire severity, feedback effects and resilience to alternative community states in forest assemblages. *Forest Ecology & Management*, **265**, 47–54.

Kool, A, & Thulin, M. (2017). A giant spurrey on a tiny island: On the phylogenetic position of *Sanctambrosia manicata* (Caryophyllaceae) and the generic circumscriptions of *Spergula, Spergularia* and *Rhodalsine. Taxon*, **66**, 615–622.

Kuchling, G, & Griffiths, O. (2012). Endoscopic imaging of gonads, sex ratios, and occurrence of intersexes in juvenile captive-bred Aldabra giant tortoises. *Chelonian Conservation & Biology*, **11**, 91–96.

Kudoh, H, & Whigham, D F. (2001). A genetic analysis of hydrologically dispersed seeds of *Hibiscus moscheutos* (Malvaceae). *American Journal of Botany*, **88**, 588–593.

Kudoh, H, Shimamura, R, Takayama, K, & Whigham, D F. (2006). Consequences of hydrochory in *Hibiscus. Plant Species Biology*, **21**, 127–133.

Kudoh, H, Takayama, K, & Kachi, N. (2013). Loss of seed buoyancy in *Hibiscus glaber* on the oceanic Bonin Islands. *Pacific Science*, **67**, 591–597.

Kuhn, T S. (1962). *The Structure of Scientific Revolutions.* Chicago: University of Chicago Press. ISBN 0-226-45808-3

Kuznetsova, A, Brockhoff, P B, & Christensen, R H B. (2017). lmerTest Package: Tests in Linear Mixed Effects Models. *Journal of Statistical Software*, **82**, 1–26, doi:10.18637/jss.v082.i13.

Lahti, D C, Johnson, N A, Ajie, B C, et al. (2009). Relaxed selection in the wild. *Trends in Ecology & Evolution*, **24**, 487–496.

Lamont, B B, & He, T. (2017). Fire-proneness as a prerequisite for the evolution of fire-adapted traits. *Trends in Plant Science*, **22**, 278–288.

Lamont, B B, He, T, & Pausas, J G. (2017). African geoxyles evolved in response to fire; frost came later. *Evolutionary Ecology*, **31**(5), 603–617.

Lawes, M J, Richardson, S J, Clarke, P J, Midgley, J J, McGlone, M S, & Bellingham, P J. (2014). Bark thickness does not explain the different susceptibility of Australian and New Zealand temperate rain forests to anthropogenic fire. *Journal of Biogeography*, **41**, 1467–1477.

Lázaro, A, & Traveset, A. (2005). Spatio-temporal variation in the pollination mode of *Buxus balearica* (Buxaceae), an ambophilous and selfing species: mainland-island comparison. *Ecography*, **28**, 640–652.

Le, M, Raxworthy, C J, McCord, W P, & Mertz, L. (2006). A molecular phylogeny of tortoises (Testudines: Testudinidae) based on mitochondrial and nuclear genes. *Molecular Phylogenetics and Evolution*, **40**, 517–531.

Lee, W G. (2001). Fifty years of Takahe conservation research and management: what have we learnt? In: *The Takahe: Fifty Years of Conservation Management and Research* (pp. 49–60). Eds: Lee, W G & Jamieson, I G. Dunedin: University of Otago Press.

Lee, W G, Wood, J R, & Rogers, G M. (2010). Legacy of avian-dominated plant-herbivore systems in New Zealand. *New Zealand Journal of Ecology*, **34**, 28–47.

Leishman, M R, & Westoby, M. (1994). The role of large seed size in shaded conditions: Experimental evidence. *Functional Ecology*, **8**, 205– 214.

Leishman, M R, Wright, I J, Moles, A T, & Westoby, M. (2000). The evolutionary ecology of seed size. In *Seeds: The Ecology of Regeneration in Plant Communities* (pp. 31–57). Eds: Fenner, M. Wallingford: CABI Publishing.

Lens, F, Davin, N, Smets, E, & del Arco, M. (2013). Insular woodiness on the Canary Islands: a remarkable case of convergent evolution. *International Journal of Plant Sciences*, **174**, 992–1013.

Lev-Yadun, S. (2016). *Defensive (Anti-herbivory) Coloration in Land Plants*. Basel: Springer Publishing.

Liston, A, Robinson, W A, Piñero, D, & Alvarez-Buylla, E R. (1999). Phylogenetics of *Pinus* (Pinaceae) based on nuclear ribosomal DNA internal transcribed spacer region sequences. *Molecular Phylogenetics & Evolution*, **11**, 95–109.

Livezey, B C. (2003). *Evolution of Flightlessness in Rails (Gruiformes: Rallidae): Phylogenetic, Ecomorphological, and Ontogenetic Perspectives*. Ornithological Monographs no. 53. Washington, DC: American Ornithologists' Union.

Lloyd, D G. (1985). Progress in understanding the natural history of New Zealand plants. *New Zealand Journal of Botany*, **23**, 707–722.

Locatelli, E, Due, R A, van den Bergh, G D, & Van Den Hoek Ostende, L W. (2012). Pleistocene survivors and Holocene extinctions: the giant rats from Liang Bua (Flores, Indonesia). *Quaternary International*, **281**, 47–57.

Lokatis, S, & Jeschke, J M. (2018). The island rule: An assessment of biases and research trends. *Journal of Biogeography*, **45**, 289–303.

Lomolino, M V. (2005). Body size evolution in insular vertebrates: Generality of the island rule. *Journal of Biogeography*, **32**, 1683–1699.

Lomolino, M V. (2010). Four Darwinian themes on the origin, evolution and preservation of island life. *Journal of Biogeography*, **37**, 985–994.

Lomolino, M V, Sax, D F, Palombo, M R, & Van Der Geer, A A. (2012). Of mice and mammoths: evaluations of causal explanations for body size evolution in insular mammals. *Journal of Biogeography*, **39**, 842–854.

Lomolino, M V, Sax, D F, Riddle, B R, & Brown, J H. (2006). The island rule and a research agenda for studying ecogeographical patterns. *Journal of Biogeography*, **33**, 1503–1510.

Lönnberg, K, & Eriksson, O. (2013). Rules of the seed size game: Contests between large-seeded and small-seeded species. *Oikos*, **122**, 1080–1084.

Lord, J M. (2004). Frugivore gape size and the evolution of fruit size and shape in Southern Hemisphere floras. *Austral Ecology*, **29**, 430–436.

Lord, J M. (2015). Patterns in floral traits and plant breeding systems on Southern Ocean Islands. *AoB Plants*, **7**, plv095.

Loyn, R H. (1997). Effects of an extensive wildfire on birds in far eastern Victoria. *Pacific Conservation Biology*, **3**, 221–234.

Lusk, C H, & Laughlin, D C. (2017). Regeneration patterns, environmental filtering and tree species coexistence in a temperate forest. *New Phytologist*, **213**, 657–668.

Lyras, G A, Dermitzakis, M D, Van der Geer, A A E, Van der Geer, S B, & De Vos, J. (2009). The origin of Homo floresiensis and its relation to evolutionary processes under isolation. *Anthropological Science*, **117**, 33–43.

Mabberley, D J. (1979). Pachycaul plants and islands. In: *Plants on Islands*. Eds: Bramwell, D. London: Academic Press.

MacArthur, R H. (1972) *Geographical Ecology: Patterns in the Distribution of Species*. Princeton University Press, Oxford.

MacArthur, R H, Diamond, J M, & Karr, J R. (1972). Density compensation in island faunas. *Ecology*, **53**, 330–342.

MacLean, W P, & Holt, R D. (1979). Distributional patterns in St. Croix Sphaerodactylus lizards: the taxon cycle in action. *Biotropica*, **11**, 189–195.

Magdalena, C. (2015). Raising the living dead: *Ramosmania rodriguesii*. *Sibbaldia: Journal of Botanic Garden Horticulture*, **8**, 63–73.

Martén-Rodríguez, S, Quesada, M, Castro, A A, Lopezaraiza-Mikel, M, & Fenster, C B. (2015). A comparison of reproductive strategies between island and mainland Caribbean *Gesneriaceae*. *Journal of Ecology*, **103**, 1190–1204.

Martin, P S, & Klein, R G. (Eds.). (1989). *Quaternary Extinctions: A Prehistoric Revolution*. Tucson: University of Arizona Press.

Martin, R D, MacLarnon, A M, Phillips, J L, Dussubieux, L, Williams, P R, & Dobyns, W B. (2006). Comment on "The brain of LB1, Homo floresiensis". *Science*, **312**, 999–999.

McCulloch, B, & Cox, G. (1992). *Moas: Lost Giants of New Zealand*. Auckland: Harper Collins.

McGlone, M S, & Webb, C J. (1981). Selective forces influencing the evolution of divaricating plants. *New Zealand Journal of Ecology*, **4**, 20–28.

McNab, B K, & Ellis, H I. (2006). Flightless rails endemic to islands have lower energy expenditures and clutch sizes than flighted rails on islands and continents. *Comparative Biochemistry and Physiology Part A: Molecular & Integrative Physiology*, **145**, 295–311.

Meiri, S. (2007). Size evolution in island lizards. *Global Ecology and Biogeography*, **16**, 702–708.

Meiri, S, Dayan, T, & Simberloff, D. (2006). The generality of the island rule reexamined. *Journal of Biogeography*, **33**, 1571–1577.

Meiri, S, Cooper, N, & Purvis, A. (2008). The island rule: made to be broken? *Proceedings of the Royal Society of London B: Biological Sciences*, **275**, 141–148.

Meiri, S, Raia, P, & Phillimore, A B. (2011). Slaying dragons: Limited evidence for unusual body size evolution on islands. *Journal of Biogeography*, **38**, 89–100.

Midway, S R, & Hodge, A M C. (2012). Carlquist revisited: History, success, and applicability of a natural history model. *Biology & Philosophy*, **27**, 497–520.

Mills, J A, & Mark, A F. (1977). Food preferences of takahe in Fiordland National Park, New Zealand, and the effect of competition from introduced red deer. *Journal of Animal Ecology*, **46**, 939–958.

Milton, S J. (1991). Plant spinescence in arid southern Africa: Does moisture mediate selection by mammals? *Oecologia*, **87**, 279–287.

Mitchell, K J, Llamas, B, Soubrier, J, et al. (2014). Ancient DNA reveals elephant birds and kiwi are sister taxa and clarifies ratite bird evolution. *Science*, **344**, 898–900.

Moles, A T, & Westoby, M. (2004). What do seedlings die from and what are the implications for evolution of seed size? *Oikos*, **106**, 193–199.

Molina-Freaner, F, Cervantes-Salas, M, Morales-Romero, D, Buchmann, S, & Fleming, T H. (2003). Does the pollinator abundance hypothesis explain geographic variation in the breeding system of *Pachycereus pringlei*? *International Journal of Plant Sciences*, **164**, 383–393.

Montgomery, S H. (2013). Primate brains, the 'island rule' and the evolution of *Homo floresiensis*. *Journal of Human Evolution*, **65**, 750–760.

Montgomery, R A, & Givnish, T J. (2008). Adaptive radiation of photosynthetic physiology in the Hawaiian lobeliads: dynamic photosynthetic responses. *Oecologia*, **155**, 455–467.

Moore, M J, Francisco-Ortega, J, Santos-Guerra, A, & Jansen, R K. (2002). Chloroplast DNA evidence for the roles of island colonization and extinction in *Tolpis* (Asteraceae: Lactuceae). *American Journal of Botany*, **89**, 518–526.

Morden, C W, Gardner, D E, & Weniger, D A. (2003). Phylogeny and biogeography of Pacific *Rubus* subgenus Idaeobatus (Rosaceae) species: investigating the origin of the endemic Hawaiian raspberry *R. macraei*. *Pacific Science*, **57**, 181–197.

Morgan-Richards, M, Trewick, S A, & Dunavan, S. (2008). When is it coevolution? The case of ground wētā and fleshy fruits in New Zealand. *New Zealand Journal of Ecology*, **32**, 108–112.

Moritz, M A, Parisien, M A, Batllori, E, et al. (2012). Climate change and disruptions to global fire activity. *Ecosphere*, **3**, 1–22.

Morwood, M J, Brown, P, Sutikna, T, et al. (2005). Further evidence for small-bodied hominins from the Late Pleistocene of Flores, Indonesia. *Nature*, **437**, 1012.

Morwood, M J, Soejono, R P, Roberts, R G, et al. (2004). Archaeology and age of a new hominin from Flores in eastern Indonesia. *Nature*, **431**, 1087.

Muenchow, G E. (1987). Is dioecy associated with fleshy fruit? *American Journal of Botany*, **74**, 287–293.

Naveh, Z. (1975). The evolutionary significance of fire in the Mediterranean region. *Vegetatio*, **29**, 199–208.

Neal, V E, & Trewick, S A. (2008). The age and origin of the Pacific Islands: A geological overview. *Philosophical Transactions of the Royal Society of London, Series B*, **363**, 3293–3308.

Newstrom, L, & Robertson, A. (2005). Progress in understanding pollination systems in New Zealand. *New Zealand Journal of Botany*, **43**, 1–59.

Nicholls, H. (2014). *The Galapagos: A Natural History*. London: Profile Books Ltd.

Niven, J E. (2007). Brains, islands and evolution: breaking all the rules. *Trends in Ecology & Evolution*, **22**, 57–59.

Noonan, J P, Coop, G, Kudaravalli, S, et al. (2006). Sequencing and analysis of Neanderthal genomic DNA. *Science*, **314**, 1113–1118.

O'Connell, D M, Monks, A, Lee, W G, Downs, T M, & Dickinson, K J. (2010). Leaf domatia: Carbon-limited indirect defence? *Oikos*, **119**(10), 1591–1600.

Obeso, J R. (1997). The induction of spinescence in European holly leaves by browsing ungulates. *Plant Ecology*, **129**, 149–156.

Olson, M E. (2003). Stem and leaf anatomy of the arborescent Cucurbitaceae *Dendrosicyos socotrana* with comments on the evolution of pachycauls from lianas. *Plant Systematics & Evolution*, **239**, 199–214.

Olson, M E, Aguirre-Hernández, R, & Rosell, J A. (2009). Universal foliage-stem scaling across environments and species in dicot trees: Plasticity, biomechanics and Corner's Rules. *Ecology Letters*, **12**, 210–219.

Owens, S J, Jackson, A, Maunder, M, Rudall, P, & Johnson, M A. (1993). The breeding system of *Ramosmania heterophylla* – Dioecy or heterostyly? *Botanical Journal of the Linnean Society*, **113**, 77–86.

Pailler, T, Humeau, L, & Thompson, J D. (1998). Distyly and heteromorphic incompatibility in oceanic island species of *Erythroxylum* (Erythroxylaceae). *Plant Systematics & Evolution*, **213**, 187–198.

Palmer, M. (2002). Testing the 'island rule' for a tenebrionid beetle (Coleoptera, Tenebrionidae). *Acta Oecologica*, **23**, 103–107.

Pannell, J R. (2015). Evolution of the mating system in colonizing plants. *Molecular Ecology*, **24**, 2018–2037.

Pannell, J R, Auld, J R, Brandvain, Y, et al. (2015). The scope of Baker's law. *New Phytologist*, **208**, 656–667.

Papadopulos, A S, Baker, W J, Crayn, D, Butlin, R K, Kynast, R G, Hutton, I, & Savolainen, V. (2011). Speciation with gene flow on Lord Howe Island. *Proceedings of the National Academy of Sciences*, **108**, 13188–13193.

Paula, S, Naulin, P I, Arce, C, Galaz, C, & Pausas, J G. (2016). Lignotubers in Mediterranean basin plants. *Plant Ecology*, **217**, 661–676.

Pausas, J G. (2015). Bark thickness and fire regime. *Functional Ecology*, **29**, 315–327.

Pausas, J G, & Keeley, J E. (2014). Evolutionary ecology of resprouting and seeding in fire-prone ecosystems. *New Phytologist*, **204**, 55–65.

Pausas, J G, & Keeley, J E. (2017). Epicormic resprouting in fire-prone ecosystems. *Trends in Plant Science*, **22**, 1008–1015.

Pausas, J G, Lamont, B B, Paula, S, Appezzato-da-Glória, B, & Fidelis, A. (2018). Unearthing belowground bud banks in fire-prone ecosystems. *New Phytologist*, **217**, 1435–1448.

Pausas, J G, & Parr, C L. (2018). Towards an understanding of the evolutionary role of fire in animals. *Evolutionary Ecology*, **32**, 113–125.

Patiño, J, Bisang, I, Hedenäs, L, et al. (2013). Baker's law and the island syndromes in bryophytes. *Journal of Ecology*, **101**, 1245–1255.

Paxinos, E E, James, H F, Olson, S L, Sorenson, M D, Jackson, J, & Fleischer, R C. (2002). mtDNA from fossils reveals a radiation of Hawaiian geese recently derived from the Canada goose (*Branta canadensis*). *Proceedings of the National Academy of Sciences*, **99**, 1399–1404.

Pelser, P B, Gravendeel, B, & van der Meijden, R. (2002). Tackling speciose genera: species composition and phylogenetic position of *Senecio* sect. Jacobaea (Asteraceae) based onplastid and nrDNA sequences. *American Journal of Botany*, **89**, 929–939.

Pérez-García, A, Vlachos, E, & Arribas, A. (2017). The last giant continental tortoise of Europe: A survivor in the Spanish Pleistocene site of Fonelas P-1. *Palaeogeography, Palaeoclimatology, Palaeoecology*, **470**, 30–39.

Perry, G L, Wilmshurst, J M, & McGlone, M S. (2014). Ecology and long-term history of fire in New Zealand. *New Zealand Journal of Ecology*, **38**, 157–176.

Pollock, M L, Lee, W G, Walker, S, & Forrester, G. (2007). Ratite and ungulate preferences for woody New Zealand plants: Influence of chemical and physical traits. *New Zealand Journal of Ecology*, **31**, 68–78.

Poole, A L, & Adams, N M. (1964). *Trees and Shrubs of New Zealand*. 3rd edition. Auckland: R E Owen, Government Printer.

Popper, K. (1959). *The Logic of Scientific Discovery*. London: Routledge.

Punzalan, D, Rodd, F H, & Hughes, K A. (2005). Perceptual processes and the maintenance of polymorphism through frequency-dependent predation. *Evolutionary Ecology*, **19**, 303–320.

Puurtinen, M, & Kaitala, V. (2002). Mate-search efficiency can determine the evolution of separate sexes and the stability of hermaphroditism in animals. *American Naturalist*, **160**, 645–660.

Randell, R A, Howarth, D G, & Morden, C W. (2004). Genetic analysis of natural hybrids between endemic and alien *Rubus* (Rosaceae) species in Hawaii. *Conservation Genetics*, **5**, 217–230.

R Core Team. (2013) *R: A Language and Environment for Statistical Computing*. Vienna: R Foundation for Statistical Computing.

Rech, A R, Dalsgaard, B, Sandel, B, et al. (2016). The macroecology of animal versus wind pollination: ecological factors are more important than historical climate stability. *Plant Ecology & Diversity*, **9**, 253–262.

Riba, M, Mayol, M, Giles, B E, et al. (2009). Darwin's wind hypothesis: Does it work for plant dispersal in fragmented habitats? *New Phytologist*, **183**, 667–677.

Rick, T C, Hofman, C A, Braje, T J, et al. (2012). Flightless ducks, giant mice and pygmy mammoths: Late Quaternary extinctions on California's Channel Islands. *World Archaeology*, **44**, 3–20.

Ricklefs, R E, & Bermingham, E. (2002). The concept of the taxon cycle in biogeography. *Global Ecology & Biogeography*, **11**, 353–361.

Ricklefs, R E, & Cox, G W. (1978). Stage of taxon cycle, habitat distribution, and population density in the avifauna of the West Indies. *American Naturalist*, 875–895.

Rickson, F R. (1977). Progressive loss of ant-related traits of *Cecropia peltata* on selected Caribbean islands. *American Journal of Botany*, **64**, 585–592.

Rico-Gray, V, & Oliveira, P S. (2007). *The Ecology and Evolution of Ant-Plant Interactions*. Chicago: University of Chicago Press.

Rigolot, E. (2004). Predicting postfire mortality of *Pinus halepensis* Mill. and *Pinus pinea* L. *Plant Ecology*, **171**, 139–151.

Rix, M, & Lewis, G. (2016). *Sophora cassiodes*. *Curtis's Botanical Magazine*, **33**, 338–346.

Rhodin, A G J, Thomson, S, Georgalis, G L, et al. (2015). Turtles and tortoises of the world during the rise and global spread of humanity: first checklist and review of extinct Pleistocene and Holocene chelonians. *Chelonian Research Monographs*, **5**, 1–66.

Robichaux, R H, Carr, G D, Liebman, M, & Pearcy, R W. (1990). Adaptive radiation of the Hawaiian silversword alliance (Compositae-Madiinae): ecological, morphological, and physiological diversity. *Annals of the Missouri Botanical Garden*, **77**, 64–72.

Rogers, B M, Soja, A J, Goulden, M L, & Randerson, J T. (2015). Influence of tree species on continental differences in boreal fires and climate feedbacks. *Nature Geoscience*, **8**, 228.

Rosell, J A. (2016). Bark thickness across the angiosperms: more than just fire. *New Phytologist*, **211**, 90–102.

Rubio de Casas, R, Willis, C G, Pearse, W D, Baskin, C C, Baskin, J M, & Cavender-Bares, J. (2017). Global biogeography of seed dormancy is determined by seasonality and seed size: a case study in the legumes. *New Phytologist*, **214**, 1527–1536.

Ryan, P G. (1987). The incidence and characteristics of plastic particles ingested by seabirds. *Marine Environmental Research*, **23**, 175–206.

Sakai A K, Wagner W L, Ferguson D M, & Herbst, D R. (1995). Origins of dioecy in the Hawaiian Flora. *Ecology*, **76**, 2517–2529.

Sakai, A K, & Weller, S G. (1999). Gender and sexual dimorphism in flowering plants: a review of terminology, biogeographic patterns, ecological correlates, and phylogenetic approaches. In *Gender and Sexual Dimorphism in Flowering Plants* (pp. 1–31). Berlin, Heidelberg: Springer.

Santana, V M, Bradstock, R A, Ooi, M K, Denham, A J, Auld, T D, & Baeza, M J. (2010). Effects of soil temperature regimes after fire on seed dormancy and germination in six Australian Fabaceae species. *Australian Journal of Botany*, **58**, 539–545.

Savolainen, V, Anstett, M C, Lexer, C, et al. (2006). Sympatric speciation in palms on an oceanic island. *Nature*, **441**, 210.

Scalon, M C, & Wright, I J. (2015). A global analysis of water and nitrogen relationships between mistletoes and their hosts: broad-scale tests of old and enduring hypotheses. *Functional Ecology*, **29**, 1114–1124.

Schaefer, H, Heibl, C, & Renner, S S. (2009). Gourds afloat: a dated phylogeny reveals an Asian origin of the gourd family (Cucurbitaceae) and numerous oversea dispersal events. *Proceedings of the Royal Society of London, B*, **276**, 843–851.

Schleuning, M, Böhning-Gaese, K, Dehling, D M, & Burns, K C. (2014). At a loss for birds: insularity increases asymmetry in seed-dispersal networks. *Global Ecology & Biogeography*, **23**, 385–394.

Schluter, D. (2000). *The Ecology of Adaptive Radiation*. Oxford: Oxford University Press.

Schoenherr, A A, Feldmeth, C R, & Emerson, M J. (2003). *Natural History of the Islands of California*. Berkeley: University of California Press.

Schueller, S K. (2007). Island-mainland difference in *Nicotiana glauca* (Solanaceae) corolla length: a product of pollinator-mediated selection? *Evolutionary Ecology*, **21**, 81–98.

Semprebon, G M, Rivals, F, Fahlke, J M, Sanders, W J, Lister, A M, & Göhlich, U B. (2016). Dietary reconstruction of pygmy mammoths from Santa Rosa Island of California. *Quaternary International*, **406**, 123–136.

Shapiro, B, Sibthorpe, D, Rambaut, A, et al. (2002). Flight of the dodo. *Science*, **295**, 1683–1683.

Shepherd, L D, & Heenan, P B. (2017a). Origins of beach-cast Sophora seeds from the Kermadec and Chatham Islands. *New Zealand Journal of Botany*, **55**, 241–248.

Shepherd, L D, & Heenan, P B. (2017b). Evidence for both long-distance dispersal and isolation in the Southern Oceans: molecular phylogeny of *Sophora* sect Edwardsia (Fabaceae). *New Zealand Journal of Botany*, **55**, 334–346.

Shepherd, L D, Lange, P J, Perrie, L R, & Heenan, P B. (2017). Chloroplast phylogeography of New Zealand *Sophora* trees (Fabaceae): extensive hybridization and widespread last glacial maximum survival. *Journal of Biogeography*, **44**, 1640–1651.

Shipley, L A. (2007). The influence of bite size on foraging at larger spatial and temporal scales by mammalian herbivores. *Oikos*, **116**, 1964–1974.

Silver, E A, Reed, D, McCaffrey, R, & Joyodiwiryo, Y. (1983). Back arc thrusting in the eastern Sunda Arc, Indonesia: A consequence of arc-continent collision. *Journal of Geophysical Research: Solid Earth*, **88**, 7429–7448.

Slikas, B. (2003). Hawaiian birds: lessons from a rediscovered avifauna. *Auk*, **120**, 953–960.

Slikas, B, Olson, S L, & Fleischer, R C. (2002). Rapid, independent evolution of flightlessness in four species of Pacific Island rails (Rallidae): an analysis based on mitochondrial sequence data. *Journal of Avian Biology*, **33**, 5–14.

Smith, J M B. (2012). Evidence for long-distance dispersal of *Sophora microphylla* to sub-Antarctic Macquarie Island. *New Zealand Journal of Botany*, **50**, 83–85.

Sorensen, A E. (1986). Seed dispersal by adhesion. *Annual Review of Ecology and Systematics*, **17**, 443–463.

Spears Jr, E E. (1987). Island and mainland pollination ecology of *Centrosema virginianum* and *Opuntia stricta*. *Journal of Ecology*, **75**, 351–362.

Steadman, D W. (2006). *Extinction and Biogeography of Tropical Pacific Birds*. Chicago: University of Chicago Press.

Stebbins, G L. (1957). Self-fertilization and population variability in the higher plants. *American Naturalist*, **91**, 337–354.

Stephens, J M C, Molan, P C, & Clarkson, B D. (2005). A review of *Leptospermum scoparium* (Myrtaceae) in New Zealand. *New Zealand Journal of Botany*, **43**, 431–449.

Stephens, S L, & Libby, W J. (2006). Anthropogenic fire and bark thickness in coastal and island pine populations from Alta and Baja California. *Journal of Biogeography*, **33**, 648–652.

Stewart, A. (1911). *A Botanical Survey of the Galápagos Islands*. Vol. 1. San Francisco: California Academy of Sciences.

Strauss, S, & Zangerl, A. (2002). Plant–insect interactions in terrestrial ecosystems. In: *Plant–Animal Interactions, an Evolutionary Approach* (pp. 77–106). Eds: Herrera, C M & Pellmyr, O. Oxford: Blackwell Publishing.

Stuessy, T F, Jakubowsky, G, Gómez, R S, Pfosser, M, Schlüter, P M, Fer, T, ... & Kato, H. (2006). Anagenetic evolution in island plants. *Journal of Biogeography*, **33**, 1259–1265.

Stuessy, T F, Crawford, D J & Ruiz, E A. (2018). Patterns in phylogeny. In: *Plants of oceanic islands: evolution, biogeography & conservation of the flora of the Juan Fernandez (Robinson Crusoe) Archipelago*. Eds: Stuessy, T F, Crawford, D J, López-Sepúlveda, P, Baeza, C M & Ruiz, E A. Cambridge, UK: Cambridge University Press.

Sugiura, S, Abe, T, & Makino, S I. (2006). Loss of extrafloral nectary on an oceanic island plant and its consequences for herbivory. *American Journal of Botany*, **93**, 491–495.

Sutikna, T, Tocheri, M W, Morwood, M J, et al. (2016). Revised stratigraphy and chronology for Homo floresiensis at Liang Bua in Indonesia. *Nature*, **532**, 366–369.

Swenson, U, & Manns, U. (2003). Phylogeny of *Pericallis* (Asteraceae): a total evidence approach reappraising the double origin of woodiness. *Taxon*, **52**, 533–548.

Swihart, R K, & Bryant, J P. (2001). Importance of biogeography and ontogeny of woody plants in winter herbivory by mammals. *Journal of Mammalogy*, **82**, 1–21.

Sykes, W R, & Godley, E J. (1968). Transoceanic dispersal in *Sophora* and other genera. *Nature*, **218**, 495–496.

Syrmos, N C. (2011). Microcephaly in ancient Greece – The Minoan microcephalus of Zakros. *Child's Nervous System*, **27**, 685–686.

Takayama, K, Ohi-Toma, T, Kudoh, H, & Kato, H. (2005). Origin and diversification of Hibiscus glaber, species endemic to the oceanic Bonin Islands, revealed by chloroplast DNA polymorphism. *Molecular Ecology*, **14**, 1059–1071.

Takayama, K, Kajita, T, Murata, J I N, & Tateishi, Y. (2006). Phylogeography and genetic structure of Hibiscus tiliaceus – Speciation of a pantropical plant with sea-drifted seeds. *Molecular Ecology*, **15**, 2871–2881.

Takayama, K, Crawford, D J, López-Sepúlveda, P, Greimler, J, & Stuessy, T F. (2018). Factors driving adaptive radiation in plants of oceanic islands: a case study from the Juan Fernández Archipelago. *Journal of Plant Research*, **131**, 469–485.

Talavera, M, Arista, M, & Ortiz, P L. (2012). Evolution of dispersal traits in a biogeographical context: A study using the heterocarpic *Rumex bucephalophorus* as a model. *Journal of Ecology*, **100**, 1194–1203.

Tapias, R, Climent, J, Pardos, J A, & Gil, L. (2004). Life histories of Mediterranean pines. *Plant Ecology*, **171**, 53–68.

Tapias, R, Gil, L, Fuentes-Utrilla, P, & Pardos, J A. (2001). Canopy seed banks in Mediterranean pines of south-eastern Spain: a comparison between *Pinus halepensis* Mill., *P. pinaster* Ait., *P. nigra* Arn. and *P. pinea* L. *Journal of Ecology*, **89**, 629–638.

Tennyson, A J D, & Martinson, P. (2006). *Extinct Birds of New Zealand*. Wellington: Te Papa Press.

Thimmappa, R, Geisler, K, Louveau, T, O'Maille, P, & Osbourn, A. (2014). Triterpene biosynthesis in plants. *Annual Review of Plant Biology*, **65**, 225–257.

Thomson, J D, & Barrett, S C. (1981). Selection for outcrossing, sexual selection, and the evolution of dioecy in plants. *American Naturalist*, **118**, 443–449.

Thorsen, M J, Dickinson, K J, & Seddon, P J. (2009). Seed dispersal systems in the New Zealand flora. *Perspectives in Plant Ecology, Evolution & Systematics*, **11**, 285–309.

Tinbergen, L. (1960). The natural control of insects in pinewoods. Factors influencing the intensity of predation by songbirds. *Archives Neerlandaises de Zoologie*, **13**, 265–343.

Traveset, A, Fernández-Palacios, J M, Kueffer, C, Bellingham, P J, Morden, C, & Drake, D R. (2016a). Introduction to the Special Issue: Advances in island plant biology since Sherwin Carlquist's Island Biology. *AoB Plants*, **8**.

Traveset, A, Tur, C, Trøjelsgaard, K, Heleno, R, Castro-Urgal, R, & Olesen, J M. (2016b). Global patterns of mainland and insular pollination networks. *Global Ecology & Biogeography*, **25**, 880–890.

Traveset, A, & Navarro, L. (2018). Plant reproductive ecology and evolution in the Mediterranean islands: State of the art. *Plant Biology*, **20**, 63–77.

Trewick, S A. (1997). Flightlessness and phylogeny amongst endemic rails (Aves: Rallidae) of the New Zealand region. *Philosophical Transactions of the Royal Society, B*, **352**, 429–446.

Tucci, S, Vohr, S H, McCoy, R C, et al. (2018). Evolutionary history and adaptation of a human pygmy population of Flores Island, Indonesia. *Science*, **361**, 511–516.

Valdebenito, H, Stuessy, T F, Crawford, D J, & Silva, M. (1992). Evolution of Erigeron (Compositae) in the Juan Fernandez Islands, Chile. *Systematic Botany*, **17**, 470–480.

Van Damme, K. (2009). Socotra Archipelago. In: *Encyclopedia of Islands* (pp. 846–851). Eds: Gillespie, R G & Clague, D A. Berkeley: University of California Press.

Van Damme, K V, & Banfield, L. (2011). Past and present human impacts on the biodiversity of Socotra Island (Yemen): implications for future conservation. *Zoology in the Middle East*, **54**, 31–88.

Van den Bergh, G D, Meijer, H J, Awe, R D, et al. (2009). The Liang Bua faunal remains: a 95 k. yr. sequence from Flores, East Indonesia. *Journal of Human Evolution*, **57**, 527–537.

Van den Hout, P J, Mathot, K J, Maas, L R, & Piersma, T. (2010). Predator escape tactics in birds: Linking ecology and aerodynamics. *Behavioral Ecology*, **21**, 16–25.

van der Geer, A, Lyras, G, de Vos, J, & Dermitzakis, M. (2010). *Evolution of Island Mammals*. Chichester: Wiley-Blackwell.

van Grouw, H, & Hume, J P. (2016). The history and morphology of Lord Howe Gallinule or Swamphen *Porphyrio albus* (Rallidae). *Bulletin of the British Ornithologists' Club*, **136**, 172–198.

Van Valen, L. (1965). Morphological variation and width of ecological niche. *American Naturalist*, **99**, 377–390.

Vazačová, K, & Münzbergová, Z. (2014). Dispersal ability of island endemic plants: What can we learn using multiple dispersal traits? *Flora*, **209**, 530–539.

Verdcourt, B. (1996). *Ramosmania rodriguesii* Rubiaceae. *Curtis's Botanical Magazine*, **13**, 204–209.

Vetter, J. (2000). Plant cyanogenic glycosides. *Toxicon*, **38**, 11–36.

Vourc'h, G, Martin, J L, Duncan, P, Escarré, J, & Clausen, T P. (2001). Defensive adaptations of *Thuja plicata* to ungulate browsing: a comparative study between mainland and island populations. *Oecologia*, **126**, 84–93.

Wagner, W L, Herbst, D R, & Sohmer, S H. (1999). *Manual of the Flowering Plants of Hawai'i*, Vols. 1 and 2 (No. Edn 2). Honolulu: University of Hawai'i and Bishop Museum Press.

Wagstaff, S J, & Breitwieser, I. (2002). Phylogenetic relationships of New Zealand Asteraceae inferred from ITS sequences. *Plant Systematics & Evolution*, **231**, 203–224.

Walker, G P L. (1990). Geology and volcanology of the Hawaiian Islands. *Pacific Sciences*, **44**, 315–347.

Wallace (1878). *Tropical Nature and other Essays*. London: Macmillan.

Wallace, A R. (1876). *The Geographical Distribution of Animals*. London: Macmillan.

Walters, D R. (2011). *Plant Defence: Warding off Attack by Pathogens, Herbivores and Parasitic Plants*. Chichester: Wiley-Blackwell.

Wang, X R, Tsumura, Y, Yoshimaru, H, Nagasaka, K, & Szmidt, A E. (1999). Phylogenetic relationships of Eurasian pines (*Pinus*, Pinaceae) based on chloroplast rbcL, matK, rpl20-rps18 spacer, and trnV intron sequences. *American Journal of Botany*, **86**, 1742–1753.

Wardle, D A, Zackrisson, O, Hörnberg, G, & Gallet, C. (1997). The influence of island area on ecosystem properties. *Science*, **277**, 1296–1299.

Ward, P S, & Branstetter, M G. (2017). The acacia ants revisited: convergent evolution and biogeographic context in an iconic ant/plant mutualism. *Proceedings of the Royal Society of London, B*, **284**, 20162569.

Watanabe, K, Shimizu, A, & Sugawara, T. (2014). Dioecy derived from distyly and pollination in *Psychotria rubra* (Rubiaceae) occurring in the Ryukyu Islands, Japan. *Plant Species Biology*, **29**, 181–191.

Watanabe, K, & Sugawara, T. (2015). Is heterostyly rare on oceanic islands? *AoB Plants*, **7**.

Watson, S J, Taylor, R S, Nimmo, D G, et al. (2012). Effects of time since fire on birds: how informative are generalized fire response curves for conservation management? *Ecological Applications*, **22**, 685–696.

Webb C J, Lloyd D G, & Delph L F. (1999). Gender dimorphism in indigenous New Zealand seed plants. *New Zealand Journal of Botany*, **37**, 119–130.

Weigelt, P, Jetz, W, & Kreft, H. (2013). Bioclimatic and physical characterization of the world's islands. *Proceedings of the National Academy of Sciences*, **110**, 15307–15312.

Weller, S G, Sakai, A K, Wagner, W L, & Herbst, D R. (1990). Evolution of dioecy in Schiedea (Caryophyllaceae: Alsinoideae) in the Hawaiian Islands: biogeographical and ecological factors. *Systematic Botany*, **15**, 266–276.

Weller, S G, Sakai, A K, Culley, T M, Campbell, D R, & Dunbar-Wallis, A K. (2006). Predicting the pathway to wind pollination: heritabilities and genetic correlations of inflorescence traits associated with wind pollination in *Schiedea salicaria* (Caryophyllaceae). *Journal of Evolutionary Biology*, **19**, 331–342.

Westoby, M. (1998). A leaf-height-seed (LHS) plant ecology strategy scheme. *Plant & Soil*, **199**, 213–227.

Westoby, M, Leishman, M, & Lord, J. (1996). Comparative ecology of seed size and dispersal. *Philosophical Transactions of the Royal Society, B*, **351**, 1309–1318.

White, P S. (1983). Corner's rules in eastern deciduous trees: allometry and its implications for the adaptive architecture of trees. *Bulletin of the Torrey Botanical Club*, **110**, 203–212.

Whittaker, R J & Fernández-Palacios, J M. (2007). *Island Biogeography: Ecology, Evolution, and Conservation*. Oxford: Oxford University Press.

Whittaker, R J, Fernández-Palacios, J M, Matthews, T J, Borregaard, M K, & Triantis, K A. (2017). Island biogeography: Taking the long view of nature's laboratories. *Science*, **357**, eaam8326.

Wilcox, C, Van Sebille, E, & Hardesty, B D. (2015). Threat of plastic pollution to seabirds is global, pervasive, and increasing. *Proceedings of the National Academy of Sciences*, **112**, 11899–11904.

Wilder, B T, & Felger, R S. (2012). Dwarf giants, guano, and isolation: vegetation and floristic diversity of San Pedro Mártir Island, Gulf of California, Mexico. *Proceedings of the San Diego Society of Natural History*, **42**, 1–24.

Wilson, E O. (1961). The nature of the taxon cycle in the Melanesian ant fauna. *American Naturalist*, **95**, 169–193.

Wilson, K-J. (2004). *Flight of the Huia*. Canterbury: Canterbury University Press.

Wilson, S L, & Kerley, G I H. (2003a). The effect of plant spinescence on the foraging efficiency of bushbuck and boergoats: browsers of similar body size. *Journal of Arid Environments*, **55**, 150–158.

Wilson, S L, & Kerley, G I. (2003b). Bite diameter selection by thicket browsers: the effect of body size and plant morphology on forage intake and quality. *Forest Ecology & Management*, **181**, 51–65.

Wiser, S K, Allen, R B, & Platt, K H. (1997). Mountain beech forest succession after a fire at Mount Thomas Forest, Canterbury, New Zealand. *New Zealand Journal of Botany*, **35**, 505–515.

Wolf, P G, Schneider, H, & Ranker, T A. (2001). Geographic distributions of homosporous ferns: does dispersal obscure evidence of vicariance? *Journal of Biogeography*, **28**, 263–270.

Worthy, T H, Mitri, M, Handley, W D, Lee, M S, Anderson, A, & Sand, C. (2016). Osteology supports a stem-galliform affinity for the giant extinct flightless bird *Sylviornis neocaledoniae* (Sylviornithidae, Galloanseres). *PLoS ONE*, **11**, e0150871.

Wright, I J, Reich, P B, Westoby, M, et al. (2004). The worldwide leaf economics spectrum. *Nature*, **428**, 821.

Wright, I J, Dong, N, Maire, V, et al. (2017). Global climatic drivers of leaf size. *Science*, **357**, 917–921.

Yamada, T, Kashiwagi, T, Sawamura, M, & Maki, M. (2010). Floral differentiation among insular and mainland populations of *Weigela coraeensis* (Caprifoliaceae). *Plant Systematics & Evolution*, **288**, 113–125.

Yamada, T, & Maki, M. (2014). Relationships between floral morphology and pollinator fauna in insular and main island populations of *Hosta longipes* (Liliaceae). *Plant Species Biology*, **29**, 117–128.

Yamada, T, Kodama, K, & Maki, M. (2014). Floral morphology and pollinator fauna characteristics of island and mainland populations of *Ligustrum ovalifolium* (Oleaceae). *Botanical journal of the Linnean Society*, **174**, 489–501.

Yonezawa, T, Segawa, T, Mori, H, et al. (2017). Phylogenomics and morphology of extinct paleognaths reveal the origin and evolution of the ratites. *Current Biology*, **27**, 68–77.

Zotz, G, Wilhelm, K, & Becker, A. (2011). Heteroblasty – a review. *The Botanical Review*, **77**, 109–151.

Index

Printed in the United States
By Bookmasters